Donkey Farming

THE AUTHOR

Dr. Pravin H Vataliya has obtained Masters of Veterinary Science degree in Animal Genetics & Breeding in 1987 from the College of Veterinary Science & Animal Husbandry, Gujarat Agricultural University, Anand. He obtained Doctorate in Animal Genetics & Breeding from Gujarat Agricultural University in 1994 from Sardar Krushinagar.

He has served as Head of the Department of Animal Genetics & Breeding at Veterinary Colleges, Anand and Junagadh. He has also served as Research Scientist & Head, Poultry Complex, Anand and Cattle Breeding Farm, Junagadh and as In charge, Principal & Dean, Veterinary College, JAU, Junagadh. He was bestowed with "Best Teacher Award" in Veterinary Science Faculty at Junagadh Agricultural University for the year 2011-12. He was deputed to Kamdhenu University and assumed the Charge of Dean Faculty of Veterinary Science at Kamdhenu University in November 2014. He was selected as Director of Extension Education at Kamdhenu University, Gandhinagar from 25th March, 2015.

Dr. Vataliya has several publications including 96 International & National publications, Eight Booklets, Two book chapters and 25 Popular articles and has delivered talks at several national seminars and symposia, trainings and workshops. He has also organized 12 national and state level seminars, workshops and several trainings for scientists and farmers.

Donkey Farming

Pravin H. Vatalilya

2022
Associated Publishing Company
A Division of
Astral International Pvt. Ltd.
New Delhi – 110 002

Cataloging in Publication Data--DK
Courtesy: D.K. Agencies (P) Ltd. <docinfo@dkagencies.com>

Vataliya, Pravin H., author.
Donkey farming / Pravin H. Vatalilya.
pages cm
Includes bibliographical references and index.
ISBN 9789354614507

1. Donkeys. 2. Animal culture. I. Title.

LCC SF361.V38 2021 | DDC 636.182 23

Published by : **Associated Publishing Company**
A Division of
Astral International Pvt. Ltd.
– ISO 9001:2015 Certified Company –
4736/23, Ansari Road, Darya Ganj,
New Delhi-110 002
Ph. 011-43549197, 23278134
E-mail: info@astralint.com
Website: www.astralint.com

Message

India is bestowed with vast animal genetic resources, both domestic and wild. The country is home to well-known breeds of cattle, buffalo, horse, sheep, goat, donkey, camel, pig, yak, mithun, fowl and other domestic birds among the domesticated species of livestock. India is also the proud home to Asiatic Lion, Bengal Tiger, Elephants and a plethora of wild animals and birds.

We are also seasonal home to visitor birds arriving India seasonally for breeding purpose owing to favourable climatic conditions especially in winter season in certain geographical areas of the country. The vast diversity of flora and fauna including the domestic and wild animals is also the integral part of the culture of the country.

Donkeys are one of such important domestic livestock species which, although was first domesticated around six thousand years ago, has unfortunately drawn less attention of scientific community. As is known, donkeys are primarily reared for draught purposes to carry the loads of goods in harsh and difficult terrain as well as in the plains. Therefore, this beasts of burden is contributing significantly to the livelihood of some rural communities involved in transport of goods and soil through donkeys as pack animals. In the Mediterranean region of the world, the domestic donkeys were domesticated for the purpose of meat and milk in addition to draught purpose. The donkeys and their hybrids with horses, i.e. mules and hinny, have also been used as pack animals for the armies.

There is a need to take care of the donkeys in extreme heat, harsh winter and during vagaries of weather and the challenges caused by adverse climatic conditions. The animals need to be protected from infectious and non-infectious diseases as well as the diseases of public health importance. Some of the diseases of the donkeys are common to other equine diseases also; therefore, the donkey population of the country also demands veterinary care when needed. Being the draught animal, the donkeys contribute significantly to cheaper transport for movement of goods and therefore, are subject to injuries. Diseases like tetanus, being a life threatening disease needs special attention. There is also need to give attention to feeds and fodder requirements of donkey and management of the donkey population under scientific housing facilities which otherwise is rarely available to such poor animals. Preventive health care, reproduction, care of young ones, adults and old animals as well as promotion of valued products such as donkey milk also needs specific attention. The donkey milk has drawn a lot of attention due to its putative medicinal values fetching high prices. There is also need to give attention to the management of breeding animals and improvement of domestic donkeys.

This book on donkey farming is aimed at providing some basic information regarding taxonomic classification of donkeys, their major uses, reproduction, management, feeds and feeding, health care and useful products from donkeys. I hope the book will prove to be of value to the readers.

Dr. P. H. Vataliya
Director of Extension Education,
Ex- Vice Chancellor
Kamdhenu University
Gandhinagar, Gujarat

Preface

Donkeys are used as traction animals in rural areas for small-scale farmers. They are an important farm animal species which adapted to the different environments, hard conditions and harsh climates. Changes in motorized agriculture and transport caused an unfavorable situation for donkeys. In the future donkey will be a rising start, so it should be defined. The purpose of this study is to define the aspect of economic importance, uncommon usages, reproduction traits, genetics, nutrition and health care.

Donkeys are very social and usually live in a group called a herd. The herd is usually lead by one jack and consists of several Jennies in the wild. Some larger herds have been found that include several males. As long as the additional males are submissive to the dominant male, everything goes smoothly. Donkeys don't seem to form strong bonds, however. In the wild, herds often break up and reform with new members on a regular basis.

Donkeys spend their days resting during the hottest part of the day. They are more active during the mornings and evenings, traveling and eating with the herd. Domestic donkeys are typically used for pack animals or for milk, according to The Global Invasive Species Database. Sometimes donkeys are trained to protect other livestock from predators. They will stomp any animal they feel is a threat.

Working donkeys, horses and mules generate vital direct disposable income that enables millions of families to access the food they need and to pay for a wide range of expenses. They also provide essential support to households' main income generation activities particularly in the agriculture sector, for example livestock and dairy production.

The present book entitled Donkey Farming contains nine chapters covering all related disciplines. These chapters include introduction to donkey farming, classification/taxonomy, major uses of donkey, reproduction, feeds and feeding, production systems, preventive health care, useful products and world wide market niche of donkey milk farms. Related terminology is given at the end for ready reference. This book is specially developed for the donkey farmers, donkey farm managers, field veterinary technocrats and others interested in draught animal farming.

I am grateful to all those people as well as various books, manuals, periodicals, magazines, journals etc. that helped in the preparation of this book. In spite of the best efforts, it is possible that some errors may have occurred into the compilation and editing of the book. Further queries, constructive suggestions and criticisms for the improvement of the book are always welcome and shall be thankfully acknowledged.

Pravin H. Vataliya

Contents

Chapter

1

Introduction to Donkey Farming

Introduction: An Overview

- A donkey is a four-legged domesticated farm animal with hooves that's related to the horse. Donkeys are known for the braying sound they make and their long ears. Donkeys, also known as burros are asses, not horses. They are shorter and generally hardier than horses. Donkeys have been used for transporting people and goods since biblical times. While donkeys have a reputation for being stubborn, they are also notoriously smart and capable of keeping themselves and their passengers away from danger.
- The words 'donkey' and 'ass' are used interchangeably to speak about the same species of animal: *Equus africanus asinus*. This is an animal that, along with zebras and horses, forms part of the horse family Equidae. Their relationship with humans has been stepped over at least 5 millennia and they are still used today as work animals, particularly in lower-income countries.
- They can grow up to 63″ tall at the withers and have large ears to both pick up sound and act as a cooling system in hotter climates. They are not endangered, but are known to suffer cruelty as they are put to work in often harsh conditions. For this reason, certain charities exist and help raise money to look after sick donkeys and help workers to ensure they are treating them humanely. Donkeys are related to horses and zebras. They are all members of the family 'equus', *i.e.* they are equines. The donkey is a descendant of the African wild ass, which is now rare in the wild and found in only a few remote parts of north-eastern Africa.

- The donkey or ass is a domesticated animal of the horse family. Donkeys have a reputation for stubbornness, but this is due to some handlers' misinterpretation of their highly developed sense of self-preservation. It is difficult to force or frighten a donkey into doing something it sees as contrary to its own best interest, as opposed to horses who are much more willing to, for example, go along a path with unsafe footing. Although formal studies of their behaviour and cognition are rather limited, donkeys appear to be quite intelligent, cautious, friendly, playful, and eager to learn. They are many times fielded with horses due to a perceived calming effect on nervous horses.
- Donkey, also called as burro, is a domestic ass belonging to the horse family, Equidae, and descended from the African wild ass (*Equus africanus*; see ass). It is known to have been used as a beast of burden since 4000 BCE. The average donkey stands 101.6 cm (40 inches) at the shoulder, but different breeds vary significantly. The Sicilian donkey reaches only about 61 cm (24 inches), while the large ass of Majorca stands at about 157.5 cm (62 inches), and the American ass has been measured to 167.6 cm (66 inches). In colour the donkey ranges from white to gray or black and usually has a dark stripe from mane to tail and a crosswise stripe on the shoulders. The mane is short and upright and the tail, with long hair just at the end, are more similar to cows than horses. The long ears are darker at the base and tip. Although, slower than horses, donkeys are surefooted and can carry heavy loads over rough terrain. The mule is a hybrid, the offspring of the mating of a male (jackass, or jack) donkey and a female horse. In some parts of the world where horses cannot easily survive or where extreme poverty prevents locals from owning horses, donkeys are the main beasts of burden and source of transportation.

History

Evolution of Donkeys

The type '*Equus*', which comprises of all existing equines, is believed to have progressed from Dinohippus, via the midway form *Plesippus*. One of the eldest kind is *Equus simplicidens*, designated as zebra-like with a donkey-shaped head. The oldest remnant to date is ~3.5 million-year-old from Idaho, USA. The type seems to have feasted rapidly into the Old World, with the likewise aged *Equus livenzovensis* recognised from Western Europe and Russia.

Molecular phylogenies designate the most current common ancestor of all contemporary equids (members of the genus *Equus*) lived ~5.6 (3.9–7.8) mya. Direct paleogenomic sequencing of a 700,000-year-old middle Pleistocene horse metapodial bone from Canada infers a more recent 4.07 million years before current date for the most recent common ancestor (MRCA) within the range of 4.0 to 4.5 Myr BP. The eldest deviations are the Asian hemiones, followed by the African zebras. All other contemporary forms comprising of the domesticated horse belong to the subgenus *E.* (*Equus*) which diverged ~4.8 (3.2–6.5) million years ago.

The descendants of the contemporary donkey are the Nubian and Somalian categories of African wild ass. Remnants of domestic donkeys dating to the fourth millennium BC have been found in Ma'adi in Lower Egypt, and it is supposed that the domestication of the donkey was accomplished long after the domestication of cattle and sheep in the seventh and eighth millennia BC. Donkeys were perhaps the first domesticated by rustic people in Nubia, and they displaced the ox as the principal pack animal of that ethos. The domestication of donkeys aided to upsurge the movement of pastoral cultures, having the benefit over ruminants of not needing time to chew their cud, and were vital in the development of long-distance trade across Egypt. In the Dynasty IV era of Egypt, between 2675 and 2565 BC, well-off affiliates of the society were known to own over 1,000 donkeys, employed in agriculture, as dairy and meat animals and as pack animals. In 2003, the crypt of either King Narmer or King Hor-Aha (two of the first Egyptian pharaohs) was unearthed and the carcasses of ten donkeys were found suppressed in a way usually used with high ranking humans. These burials show the prominence of donkeys to the early Egyptian state and its ruler.

The first donkeys came to the Americas on ships of the Second Voyage of Christopher Columbus, and were landed at Hispaniola in 1495. The first to reach North America may have been two animals taken to Mexico by Juan de Zumárraga, the first bishop of Mexico, who attained there on 6 December 1528, while the first donkeys to reach what is now the United States may have crossed the Rio Grande with Juan de Oñate in April 1598. From that time on they feast northward, finding use in missions and mines. Donkeys were recognised as contemporary in what today is Arizona in 1679. By the Gold Rush years of the 19th century, the burro was the **beast of burden** of choice of early prospectors in the western United States. With the end of the placer mining boom, many of them bolted or were uninhibited, and a feral population recognised itself.

The Muscat or Yemen ass was established in Arabia. By the second millennium BC, the donkey was transported to Europe, probably at the similar time as viticulture was presented, as the donkey is related with the Syrian god of wine, Dionysus. Greeks spread both of these to many of their colonies, including those in what are now Italy, France and Spain; Romans detached them throughout their empire.

Difference between a Donkey and a Mule

Though one cannot find much difference between a donkey and mule, they have anatomical and physiological differences. People often find it hard to differentiate between the two — Donkey and Mule "as they are almost similar in their looks and ways. Donkey is a domesticated hoofed mammal, which is related to the horse. But a donkey is smaller than a horse with long ears, short hair on the tail, short mane and a dark stripe along the back. A mule is a hybrid animal from crossing a jack (male donkey) and a mare (female horse).

A mule is the offspring of a male donkey (called a jack) and a female horse (called a mare). Mules inherit desirable qualities from both donkeys and horses; from the horse they inherit strength and stamina, and like donkeys, mules are generally patient, sure-footed, intelligent and have an even temper. Mules are sterile but female mules are known to produce milk.

Put simply, a donkey is the descendant of the African wild ass (of which there are only about 500 left in the wild today). Donkeys are a different species than a horse, but in the same family. They were originally bred in Egypt or Mesopotamia around 5,000 years ago.

Mules are a cross between a female horse and a male donkey or "jack" (hence the word "jackass"). In contrast, a male horse and a female donkey bred together produces a "hinny"; the name coming from the fact that a female donkey is called a "jenny", bred with a horse. The offspring of this breeding is very similar to a mule, as you might expect, though generally slightly smaller, which is thought to be because of the smaller womb in a donkey over a horse.

Status of Donkeys in India

In India, majority of the donkey population is owned by socially and economically deprived landless people and these animals serve as the source of livelihood for their masters and play an important role in their socio-economic status also. The contribution of donkeys has been enormous in agriculture, herding and carrying materials with migratory herds of cattle, sheep and goats, transportation of agricultural produce, house hold material and provision as well as material for building construction. Although donkeys support the rural livelihood and low income poor farmer's families by providing economy at minimal maintenance cost, they are not yet considered as farm animals. Donkeys are neglected and rarely studied. Due to their indiscriminate use without any proper feed and fodder, health status *etc.*, these animals are also known as beast of burden. Donkeys are amongst the eldest domesticated animals. Even now in numerous fragments of this world they carry weights and are used for riding and for many families they are the only means to earn a living. So, it is in the South of India. Unfortunately, the donkeys are often kept under bad conditions. Whether construction materials for a new house or sand dragged out of the river bed to be transported, much of the work could not be done without the help of these small greyish equines. But many donkeys suffer from bad husbandry conditions: Harnesses and saddles that do not fit properly cause painful injuries. As most of the livestock owners are poor they do not have the means to care for the animals adequately. Additionally, they do not own land on which to keep the donkeys, thus the donkeys move freely and are often hit by cars in the wild traffic on India's roads. Hardly any donkey ever reaches their natural lifespan of 30 years under these conditions and only a few live to be more than three or four years old.

As per 19th livestock Census (2012), total equine population in India was about 1.14 million and donkeys constituted 28 per cent of the total equines. Total donkey population in the country has decreased by 27.22 per cent between 18th and 19th Livestock Census which is higher than the overall decrease (9.6 per cent) in equine population. This decrease in population might be due to their decreased working utility, as alternate economical sources of mechanization are available. Donkeys are scattered in almost every state in India and their statewise population is: Rajasthan ($0.81x10^5$); Uttar Pradesh ($0.56x10^5$), Gujarat ($0.38x10^5$), Maharashtra ($0.29x10^5$), Bihar ($0.21x10^5$), Jammu and Kashmir ($0.17x10^5$), Karnataka ($0.16x10^5$), Madhya Pradesh ($0.15x10^5$), Andhra Pradesh ($0.13x10^5$) and so on (19th Livestock Census 2012, Govt. of India). In some geographic areas with restricted access, donkeys have adapted themselves to those particular agro-climatic conditions also.

Important

The total population of Mules in the country is Eighty Four Thousand in 2019, decreased by 57.1 per cent over previous Census. The total population of Donkeys in the country is 1.2 Lakhs in 2019, decreased by 61.23 per cent over previous Census.

Characteristics of Donkey

Following are some of the chief characteristics of donkeys:

- ☆ Donkeys have an array of different habitats and geographic locations due to their domestication. The ancient asses, however, were discovered in southern and central Asia and northern Africa. In these areas they are well adapted to live in rocky highlands and low land deserts and prefer to live in warmer, dry climates.
- ☆ Wild donkeys are only found in deserts and savannahs in northern Africa from Morocco to Somalia, in the Arabian Peninsula and in the Middle East.
- ☆ Domesticated donkeys, on the other hand, are found all over the world, but prefer dry, warm areas. The Abyssinian donkey, for example, is bred in Ethiopia, according to Oklahoma State University, while the Anatolia donkey is bred in Turkey.
- ☆ One species, the kiang or Tibetan wild ass is found in China, northern parts of Pakistan, India, Nepal and Bhutan.

Interesting Fact

Some species of donkeys, especially African and Asian donkeys, are very territorial. They will guard large territories of their habitat for years. In fact, some donkeys will only mate after they establish a territory.

The Food and Agriculture Organization of the United Nations estimated more than 40 million donkeys lived worldwide in 2011, three-quarters of them in Africa and Asia. Countries with the most donkeys include China, Ethiopia, Pakistan, Egypt and Mexico. A donkey's living conditions are usually determined by his relationship with the people who share his habitat.

Wild Donkeys

The wild African ass, from which domestic donkeys originated, survives in the countries making up the Horn of Africa, according to Ronald M. Nowak, editor of "Walker's Mammals of the World." Wild donkeys living elsewhere in the world are not actually wild donkeys. They are feral domestic donkeys, often called burros. In the United States, an estimated 5,800 burros lived in five Western states in 2012, according to the Bureau of Land Management. These burros roam free on BLM-managed rangeland in Arizona, California, Nevada, Oregon and Utah.

Working Donkeys

Donkeys have served as pack and draft animals for about 5,000 years. Today, millions of people living in developing countries rely on donkeys to till fields, transport produce and other products to market, carry water, pull carts and work in mills. Such donkeys live wherever is most convenient for their owners. Some live on farms in corrals, and some live in urban environments — even inside houses. Donkeys work in developed countries, too. They are highly affectionate and patient. For this reason, donkeys serve in equine therapy programs for disabled riders. Donkeys sometimes serve as foal or stable companions because of their calming effect on horses. These donkeys typically live in stables and spend time out in pastures during the day.

Pet Donkeys

Donkeys are lovable, intelligent animals that aren't necessarily all workers. Some are kept as pets. They may live in a stable at night and a pasture during the day to graze; others may live in a pasture day and night. It is important that donkeys have access to some type of shelter for protection from wind and rain. A sturdy lean-to will do. Donkeys originate from warm climates; in a region with ice and snow, donkeys need a warm stable or barn.

Rescued Donkeys

Hundreds of donkeys require rescuing each year. Some are abandoned, others abused and seized. Some donkeys are surrendered to rescues by owners unable to care for them any longer. In the United States, some are feral donkeys captured by the Bureau of Land Management and placed for adoption to control protected herd populations. These rescued donkeys live in animal sanctuaries operated by non-profit animal rescue groups. The size and scope of these sanctuaries vary greatly from a few donkeys to hundreds. Understanding the needs of donkeys, rescue groups provide shelter and pasture acreage for grazing. Sturdy fences made of wood planks or wood posts and wire enclose the pastures and keep the donkeys safe.

Habits

Male donkeys are called jacks and females are called jennets or jennies. Donkeys are very social and usually live in a group called a herd. The herd is usually lead by one jack and consists of several jennies in the wild. Some larger herds have been found that include several males. As long as the additional males are submissive to the dominant male, everything goes smoothly. Donkeys don't seem to form strong bonds. In the wild, herds often break up and reform with new members on a regular basis.

Donkeys spend their days resting during the hottest part of the day. They are more active during the mornings and evenings, travelling and eating with the herd. Donkeys are extremely stoical, and show their pain or fear much less than horses. They are so stoical that even in cases of severe illness often the only symptom to alert the handler is very minor changes of behaviour. Donkeys are less expressive in their behaviour when terrified and have a reduced flight mechanism so they tend not to panic as easily as horses. This can lead to a misreading of their emotional states, and can, as a result, make them more difficult to work with than horses. The fearful horse clearly displays their fear, the fearful donkey may simply raise himself up a couple of inches in height, his eyes may widen slightly and may just show what appears to be increased interest in the situation. It is equally as easy to misdiagnose the bargey donkey, as being aggressive or disrespectful when in fact this behaviour is commonly found in donkeys that are actually fearful of their situation.

Domestic donkeys are typically used for pack animals or for milk, according to The Global Invasive Species Database. Sometimes donkeys are trained to protect other livestock from predators. They will stomp any animal they feel is a threat. Male donkeys can be aggressive with each other, often fighting to the point of drawing blood. Their stoic nature and increased fight mechanism probably accounts for the reason that 'natural horsemanship' methods are largely unsuccessful with donkeys and mules. Donkeys are not easily forced against their will and it is very difficult to make them perform. Donkeys are not difficult to train if trained using the science of behaviour. Donkeys respond very well to the use of positive reinforcement and can be clicker trained extremely well if the trainer is experienced in its use. A donkey with a fear or phobia will respond well to systematic desensitization and counter conditioning training.

Donkeys are from origin an animal which should always be ready to flight from predators, and therefore their instincts for fighting is still a part of their behaviour. They are more likely to take up a fight than horses, and they are quite territorial animals. As domesticated animals, they sometimes have difficulties even in tolerating animals of other species within their territory. In nature, they would live in small flocks. They are sensitive to changes in the environment, and as domestic animals, this also means that they are sensitive to changes in the way in which humans behave and treat them. Their behaviour often reflects the behaviour of the humans who interact with them. All changes should, therefore, be introduced slowly and gradually. They have very good memories. They want to be stimulated, and a boring environment can create behavioural problems. They are very sensitive to pain and pain should always be considered, if they behave strangely or refuse to collaborate.

Donkeys can learn quite easily, especially things which are close to their natural behaviour, and they can be slow learners when they are asked to do something unnatural like lifting feet for check and cleaning. They are very fond of being scratched, because it reflects their mutual grooming behaviour in the flock, so that is a good reward in a learning process. Sometimes the mutual relationship between donkeys and humans make one doubting who is training who, because the donkey seems to train humans in giving them rewards or scratching them at the right spots.

Donkeys are tolerant of hot, arid environments. They are strong animals and can work hard and carry burdens. They are often claimed to be stubborn, but they are really just very good survivors and that is important to understand when working with them: most of their behaviour is motivated by their will to survive, and therefore avoid everything which they find difficult, frightening or painful.

Housing

The housing of donkeys can be kept very basic. Depending on the climate and season, a small shelter is sufficient. It should have, at the very least, a roof. It should, preferably also have three closed sides with the open side facing away from the prevailing wind directions. There must be enough space to lie down and the floor should not be damp or cold. Barbed wire enclosures should not be used for donkeys. Many donkeys will try to get through or jump over and injure themselves.

Modern donkey farms comprise of sheltered stables with nutritional feed supplements, grazing fields and unobstructed access to water. These farms deliver amenities and management conditions for improving welfare of donkeys such as proper ventilation, lighting and bedding materials and emergency precautions against hazards such as fire extinguishers and precautions from injurious objects. With proper adjustments, standards boxes for horses can be adopted: The recommended box sizes are 3.66m x 3.66m for very large breeds and 3.05m x3.05m for large breeds and 4mx4m for expectant Jennies.

Donkeys have evolved to travel long distances in search of sparse vegetation. This search for food, and the environment in which the donkeys lived, kept them mentally stimulated, fit and lean. The domesticated donkey lives in a restricted environment. As a result of these restrictions, domestication can cause donkeys many problems ranging from overfeeding, boredom and poor social interaction.

One of the most important elements provided by the environment is mental stimulation. Without this stimulation, problem behaviours develop; these behaviours are often designed to relieve the boredom of domestication. Insufficient space increases the competition with herd mates for food, water, shelter and personal space; this in turn increases stress which can affect the temperament of the donkey. Owners and handlers become a part of the environment through interacting with a donkey. Changes in the behaviour of people can change the behaviour of donkeys. Any changes in the environment, herd dynamics or routine can change the behaviour of a donkey.

A donkey house can be constructed from 6 poles (one at each corner, one in the centre at the rear of the shelter, two for the door). The sides can be made of off cuts or, if these are not available, sacking can be fixed to the poles to afford shelter from the wind. The door should be at least 6 foot (1.8 m) wide. The stable needs to be at least 6 foot (1.8 m) high, and the roof should be pitched to allow water to drain. It is advisable to either gutter the roof and harvest the water or dig a ditch at the back of the shelter, in order that any water spillage is drained away from the stable.

Environmental changes should be made slowly and planned carefully to allow the donkey to adjust to changes without becoming stressed. Environmental enrichment is vital for domesticated donkeys. Donkeys display emotions through sounds and actions. A paddock is the best night accommodation for a donkey. A paddock, however, allows some grazing, although a donkey will not eat grass that has grown on manure or even urine. As early mornings and late afternoons are important times for donkeys to be grazing, if they are kept in a kraal they must spend the minimum amount of time in it.

Chapter 2

Classification/Taxonomy

Donkeys were first domesticated around 5,000 years ago as beasts of burden and companions, most likely in Egypt or Mesopotamia. There are about 41 million donkeys in the world today; China has the most with 11 million, followed by Pakistan, Ethiopia, and Mexico.

Donkeys vary considerably in size, depending on breed and management. The height at the withers ranges from 7.3 hands (31 inches or 79 cm) to 15.3 hands (63 inches or 160 cm), and they weigh from 80 to 480 kg (180-1,060 lb.). Working

Zebras and donkeys can breed together, creating a "zonkey". Generally, such breeding takes place between female donkeys and male zebras as domestic female zebras' breeding schedules tend to be booked for creating purebred zebras, unlike female donkeys who are much more numerous.

According to the International Commission Zoological Nomenclature, the scientific name for a donkey is Equus africanus asinus.

Today there are around 41 million donkeys in the world with about 185 different breeds. The largest population of donkeys can be found in China, which has about 11 million.

The U.S. army still uses mules to carry supplies in Afghanistan in places where the terrain in particularly rugged.

While mules are almost always sterile, they have been successfully cloned, the first of which was in 2003 at the University of Idaho in conjunction with Utah State University.

donkeys in the poorest countries have a life expectancy of 12-15 years; and in more prosperous countries, they may have a lifespan of 30-50 years. Many donkey breeds may be associated with different ecologic and environmental conditions. Donkeys have shown a wide range of adaptations under different environmental and climatic conditions. Two main types of donkey are worldwide recognized: the common African donkey (with a variety of coat colors and a typical stripe on the shoulder), less heavier than the Poitou donkey, generally dark brown, long haired and stouter animal, dedicated to temperate environment and used for centuries to produce mules.

***Source*: Accessed from http: //www.ilyke.net/**

There are over 189 donkey breeds in the world as listed by The Domestic Animal Diversity Information System (DAD-IS) of the FAO in June 2011. These breeds vary in body weights and sizes ranging from 80-480 kg and heights at withers measuring from 75-160 cm. Some of the breeds worth mentioning include the Abkhazskaya or Abyssinian donkey (Ethiopia), American mammoth Jack (derived from selective crossings of the Malteses, Poitou, Majorcan, Andulusian and Catalonian breeds), the Poitou ass (bred for the production of Mules) and Âne de Provence (France), Âne du Miankala (Mali), Asino dell'Asinara (Italy), Criollo (Cuba), Asno Americana (Chile), Asno Andaluz or Andalusian donkey (Spain), Barockesel referred to as the Austrian-Hungarian White Donkey, Biyang (China), Black (Jordan), Dongolawl (Sudan), Tswana (Botswana), Hassawi (Egypt), Biyang (China), Bourik (Haiti), Burro Kentucky (Hundorass, El Salvador, Brazil).

Following is the list of breeds of domestic donkey which is based on country reports to the international DAD-IS database:

Breed Name English/(native language in DAD-IS)	*Country*	*Notes*	*Image*
Abyssinian donkey	Ethiopia		
Algerian	Algeria	Used as a draught or meat animal. Chestnut or grey in colour.	
American Mammoth Jack	USA	Also called "Mammoth Jackstock" or "Mammoth donkey"; developed from a blend of Poitou, Andalusian, Majorcan, and Catalan breeds	
Anatolian	Turkey		
Bourbonnais donkey(Âne du Bourbonnais)	France		

Breed Name English/(native language in DAD-IS)	*Country*	*Notes*	*Image*
Cotentin donkey (Âne du Cotentin)	France		
Norman donkey (Âne Normand)	France		
Ânewallon	Belgium		
Asinara donkey (Asinodell' Asinara)	Italy	Critically endangered; indigenous to the Isola dell'Asinara, Sardinia	

Breed Name English/(native language in DAD-IS)	***Country***	***Notes***	***Image***
Amiatina (Asinodell' Amiata)	Italy	Monte Amiata, province of Grosseto, Tuscany	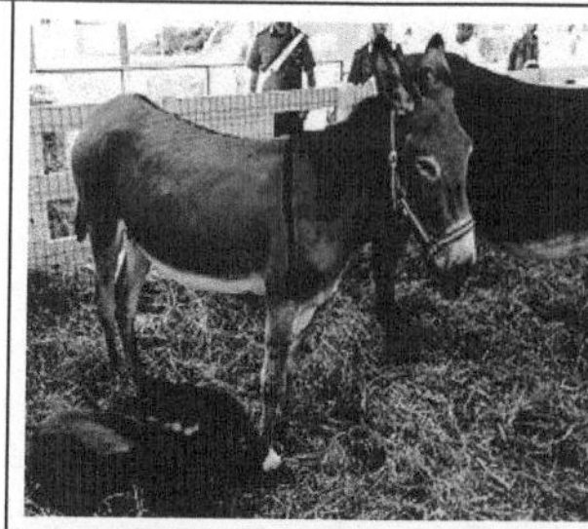
Martina Franca donkey (Asino di Martina Franca)	Italy	Martina Franca and neighbouring areas in south-east Murgia, in the Metropolitan City of Bari and the provinces of Taranto and Brindisi in Puglia	
Grigio Siciliano (Asino Grigio Siciliano)	Italy	Sicily	
Romagnolo donkey (Asino Romagnolo)	Italy	from Emilia-Romagna	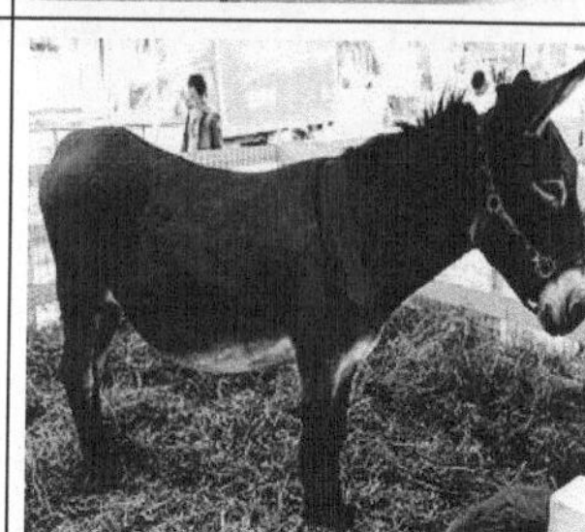

Breed Name English/(native language in DAD-IS)	***Country***	***Notes***	***Image***
Sardinian donkey (AsinoSardo)	Italy	from Sardinia	
Asno	Chile		
AsnoBalear	Spain		
Asno Criollo	Venezuela		
Asno de las Encartaciones	Spain	Basque Country	
Sperki breed	Pakistan		

Breed Name English/(native language in DAD-IS)	**Country**	**Notes**	**Image**
Australian donkey	Australia		
Barockesel/Austrian-Hungarian white Donkey	Austria		
Azerbaidzhanskaya	Azerbaijan		
Poitou donkey (Baudet de Poitou), Poitevin	France	The Poitou Donkey breed was developed for the sole purpose of producing mules. It is a large donkey breed with a very long shaggy coat and no dorsal stripe	
Bulgarian donkey	Bulgaria		

Breed Name English/(native language in DAD-IS)	***Country***	***Notes***	***Image***
Burro	Mexico, Nicaragua, USA	Small donkey of Mexico and the USA seen in both domesticated and feral states; some feral burros in the western United States are protected by federal law	
Miranda donkey (Burro do Miranda)	Portugal	Endangered status; also called Burro do Planalto Mirandes, Mirandes, Raçaasinina de Miranda, or Transmontano	
Burro Kentucky	El Salvador, Honduras		
Catalan donkey (Catalana)	Spain		
Comune	Albania		

Breed Name English/(native language in DAD-IS)	***Country***	***Notes***	***Image***
Corsican donkey	Corsica		
Criollo	Peru		
Cyprus donkey	Cyprus		
Damascus	Israel, Syrian Arab Republic		
Balkan donkey (Domacibal-kanski-magarac)	Serbia, Monte-negro		

Breed Name English/(native language in DAD-IS)	Country	Notes	Image
Irish donkey	Ireland		
Egyptian	Egypt		
Ezel	Netherlands		
Berry Black (Grand Noir du Berry)	France	Former province of Berry; possibly derived from Catalan	
Maltese donkey (Hmar Malti)	Malta		

Breed Name English/(native language in DAD-IS)	***Country***	***Notes***	***Image***
Indian	India		
Istarski-magarac	Croatia	Istria	
Italian	Serbia		
Jordanian donkey	Jordan		
Karakaçan	Turkey		

Breed Name English/(native language in DAD-IS)	***Country***	***Notes***	***Image***
Kassala	Eritrea		
Kirgizskaya	Kyrgyzstan		
Magyar parlagiszamár	Hungary		
Masai	Kenya, Tanzania		
Merzifon	Turkey		
Miniature donkey	USA		

Breed Name English/(native language in DAD-IS)	Country	Notes	Image
Moroccan	Morocco		
Muscat	Tanzania		
Native of North Africa	Mali, Mauritania, Senegal		
Ponui donkey	New Zealand		
Primorskodinarski-magarac	Croatia		

Breed Name English/(native language in DAD-IS)	***Country***	***Notes***	***Image***
Provence donkey(Âne de Provence)	France	Provence	
Puttalam Buruwa	Sri Lanka		
Pyrenean donkey(Âne des Pyrénées)	France	Aquitaine, Midi Pyrénées, Languedoc Roussillon; also called Gascon; similar to Catalan but smaller	
Qubressy	Jordan		
Romanian donkey	Romania		

Breed Name English/(native language in DAD-IS)	***Country***	***Notes***	***Image***
Sjevernoja-dranskimagarac	Croatia		
Somali	Djibouti, Kenya, Somalia, Yemen		
Standard	USA		
Thuringian Forest donkey	Germany		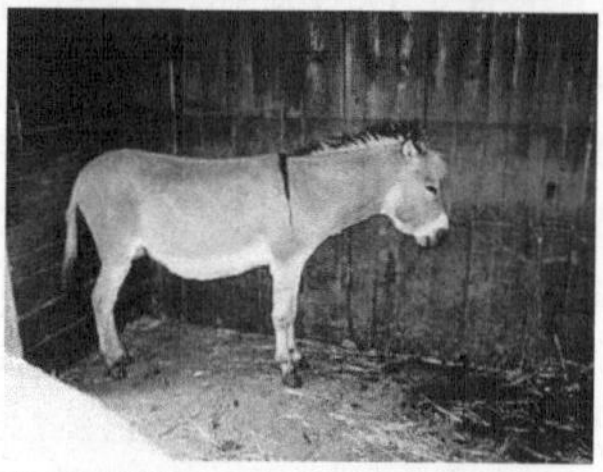
Tadzhikskaya	Tajikistan		

Breed Name English/(native language in DAD-IS)	*Country*	*Notes*	*Image*
Tunisian	Tunisia		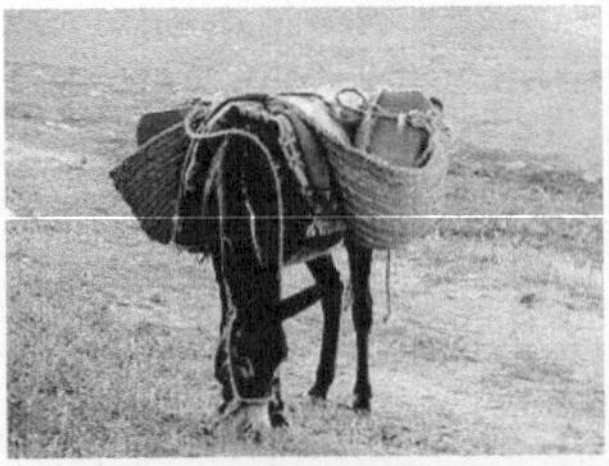
Xinjiang	China		
Zamorano-Leonés	Spain		

Classification/Taxonomy

According to Marshall and Weissbrod (2011), the subtleties of wild ass mating systems, based on short-lived relations that happen when females move through male terrains, influence donkey breeding in the domesticated environment. Some rural women stressed their concern with the belligerent behaviour of jacks during mating. Even when they wanted to keep a female from breeding with an especially aggressive jack, women said that they found it unbearable to keep the male away. They also noted that estrus Jennies might go amiss without warning in search of males. They are often lost this way, and a number of cases in which nomadic females have been documented, as well as males, were cared over a long term by women in distant settlements.

Here is the taxonomy of donkeys, according to the Integrated Taxonomic Information System (ITIS):

Kingdom: Animalia

Subkingdom: Bilateria

Infrakingdom: Deuterostomia

Phylum: Chordata

Subphylum: Vertebrata

Infraphylum: Gnathostomata

Superclass: Tetrapoda

Class: Mammalia

Subclass: Theria

Infraclass: Eutheria

Order: Perissodactyla

Family: Equidae

Genus: Equus

Species

Equus asinus — burro, donkey, ass

Equus hemippus — Syrian wild ass, achdari

Equus khur — Indian wild ass, khur

Equus africanus — African wild ass

Equus africanus africanus — Nubian wild ass

Equus africanus somaliensis — Somali wild ass

Equus hemionus — Asiatic wild ass, onager, kulan

Equus hemionus hemionus — Mongolian wild ass

Equus hemionus kulan — kulan, Turkmenistanionager

Equus hemionus onager — Iranian onager

Equus kiang — Tibetan wild ass, kiang

Equus kiang holdereri — Eastern kiang

Equus kiang kiang — Western kiang

Equus kiang polyodon — Southern kiang

Fun Fact

If a donkey becomes feral it will search out a habitat that is warm and dry. When horses are scared, they often will jump or bolt. When asses are scared, they will freeze and investigate whatever scared them, which has given them the reputation of being stubborn. Characteristically, nevertheless, donkeys are very obedient.

Chapter 3

Major Uses of Donkey

Donkeys have been good partners for mankind over several millenniums. Descended from the African wild ass, the donkey has been domesticated for approximately 5,000 years. It's not an exaggeration to state that the lowly donkey helped create civilization. They still serve as beasts of burden, draft animals and basic transportation in less developed parts of the world. In modern countries, donkeys work as riding and driving animals, livestock guardians and companions to larger equines. A major role for the donkey is cherished pet — a part at which he excels.

The phrase 'beasts of burden' describes their utility as pack animals and in many parts of the world they play a substantial economic and social role in the transport of water, building materials, relief supplies, animal feed and other critical supplies. In recent years, the demand for diverse products of donkey origin, including meat, milk and skins, has escalated rapidly and various institutions are striving to prevent donkeys suffering in the process by working in partnership with key players in government, industry and the veterinary community to promote the highest international standards of donkey welfare.

Donkey cart ambulances are an increasing trend in Africa where they are used for transporting sick people to hospital. Donkeys are particularly suited to this role because they are easy to keep and do not show fatigue. In recent years, donkey-related research is increasing with the goal of using this species to further improve human health and provide greater social benefit.

Mule Production

Without donkeys, there would be no mules. The mule is the result of a mating between a donkey jack and a horse mare. Mules are sterile and can't reproduce on their own. Less often, a horse stallion and donkey jenny are crossed, producing a hinny. While it's hard to tell the difference between a mule and a hinny, the latter is usually more horse-like in appearance.

Livestock Guardian

Donkeys have a natural aversion to dogs and other canids, including foxes and coyotes. They go after canines with their hooves and teeth. That's not a good situation for your doggy companions, but it makes the donkey a useful guardian for sheep and other livestock. You can train your donkey to accept your dog, as well as train your dog to stay out of the donkey's way. A donkey will also bray like crazy, alerting you to potential intruders. For best results, obtain a recently weaned donkey and let him grow up with your flock or herd. Stick with jennies or geldings — jacks can behave roughly with smaller livestock. Avoid miniature donkeys, as they are too small for the role of livestock guardian.

Transport

Donkeys are used like taxis to transport people around cities, towns and rural areas, for example helping people to travel to school, for work, shopping and hospital visits.

Farming and Agriculture

Whether it's ploughing, harvesting, transporting crops or water, donkeys are often used for farming across developing countries.

Commercial Work

Donkeys pull carts transporting materials, collect waste and deliver supplies. Some agencies treat donkeys keeping some of the poorest societies in the world moving, such as in Mali and Mauritania.

Tourism

In Morocco, some agencies see many donkeys used as pack animals to transport goods and equipment for tourists in harsh rural or mountainous areas.

Domestic Work

Whether it's collecting water or firewood, working donkeys can be used for domestic work instead of or in addition to income generating work. Often the domestic chores a donkey assists with means that children are able to go to school instead of having to complete these vital tasks unaided. This can be seen in Botswana.

Riding Therapy Animals

Donkeys have an affinity with people and specially trained donkeys are often used for riding therapy for children with special or complex needs. This is as true in the UK as it is in the centres of Morocco and Mali, where donkeys are cared for and trained by our expert staff.

Riding and Driving

Donkeys come in a wide range of sizes. The miniature donkey stands 36 inches tall at the withers and under, the standard donkey ranges from 48 to 54 inches, and the mammoth usually stands 56 inches or more. They generally make safe riding mounts for kids and beginners, especially since fast isn't one of their primary speeds. The American Donkey and Mule Association notes that some donkeys are even gaited, making for a smoother ride.

Donkeys make can be considered as safe, reliable driving animals when properly trained. Your donkey can take you around the neighbourhood for an afternoon drive, or you can use them to plough your vegetable garden.

In Warfare

During World War I, John Simpson Kirkpatrick, a British stretcher carrier helping with the Australian and New Zealand Army Corps, and Richard Alexander "Dick" Henderson of the New Zealand Medical Corps used donkeys to release injured soldiers from the battlefield at Gallipoli. According to British food writer Matthew Fort, donkeys were used in the Italian Army. The Mountain Fusiliers each had a donkey to carry their gear, and in dangerous conditions the animal could be eaten. Donkeys have also been used to carry explosives in battles that include the war in Afghanistan and others.

Chapter 4

Reproduction

According to Marshall and Weissbrod (2011), the dynamics of wild ass mating systems, based on short-lived associations that occur when females move through male territories, influence donkey breeding in the domesticated environment. Maasai women stressed their concern with the aggressive behaviour of jacks during mating. Even when they wanted to keep a female from breeding with an especially aggressive jack, women said that they found it impossible to keep the male away. They also

Ancient Egypt High Donkey Production Systems

The domestication of donkeys served to increase the mobility of pastoral cultures, having the advantage over ruminants of not needing time to chew their cud, and was vital in the development of long-distance trade across Egypt. In the Dynasty IV era of Egypt, between 2675 and 2565 BC, wealthy members of society were known to own over 1,000 donkeys, employed in agriculture, as dairy and meat animals and as pack animals. By the end of the fourth millennium BC, the donkey had spread to Southwest Asia, and the main breeding center had shifted to Mesopotamia by 1800 BC. The breeding of large, white riding asses made Damascus famous; while Syrian breeders developed at least three other breeds, including one preferred by women for its easy gait. The Muscat or Yemen ass was developed in Arabia. By the second millennium BC, the donkey was brought to Europe, possibly at the same time as viticulture was introduced, as the donkey is associated with the Syrian God of wine, Dionysus. Greeks spread both of these to many of their colonies, including those in what are now Italy, France and Spain; Romans dispersed them throughout their empire.

noted that estrus Jennies might go astray without warning in search of males. They are often lost this way, and a number of cases in which wandering females has been documented, as well as males, were cared over a long term by women in distant settlements.

Humans have 23 pairs of chromosomes, which enable your genetic information to be passed on to your children. When human eggs and sperm collide, you have 23 chromosomes from each parent. So your dad's chromosome No. 1 lines up with mom's No. 1, No. 2 lines up with No. 2, *etc.* Each pair is nearly identical, so when egg and sperm merge, the pairs usually combine smoothly to create another healthy baby that eventually will be capable of carrying on the species.

But combined with the dissimilarities in the horse-donkey chromosomes, the end result is usually — but not always — sterile offspring. While there are no recorded cases of fertile mule stallions, there have been a few dozen cases of mule mares giving birth after mating with a horse or donkey. In fact, the ancient Greek historian Herodotus saw a mule giving birth as an ill omen for Xerxes' invasion of Greece in 480 B.C.:

"There happened also a portent of another kind while he was still at Sardis — a mule brought forth young and gave forth a mule."

And, in 2002, the BBC ran a story about how a 14-year-old mother mule in Morocco surprised everyone by giving birth.

"The mule's aged owner did not realize the mule was pregnant and rode her (12 miles) to market the day before the birth," according to the report. Soon, streams of people were flooding the woman's farm with gifts to celebrate the miracle.

"Miracle" is probably an apt description. According to the BBC report, only 60 cases of mules giving birth were recorded from 1527 to 2002, nearly 500 years. In recent times, mules produced a filly in China in 2001 and colts to mules in Morocco in 2002 and Colorado in 2007. However, according to the American Donkey and Mule Society, only one hinny mare has ever been known to give birth, in China in 1981. On the other hand, male mules and hinnies apparently shoot nothing but blanks when it comes to producing offspring. While different species of the Equidae family can interbreed, offspring are almost always sterile. Nonetheless, horse/donkey hybrids are popular for their durability and vigor. A mule is the offspring of a jack (male donkey) and a mare (female horse). The much rarer successful mating of a male horse and a female donkey produces a hinny. Mules were common in East Africa until the 70's since when they are hardly ever seen. They were used widely as pack animals during the war and on safari. The mule is, generally, a much bigger animal than the East Africa donkey. They are usually a dark brown and their ears are much longer than a donkey's. Their size makes them stronger than a donkey, but they are not as obliging as a donkey.

Jennies are pregnant for approximately 11 months, and usually give birth to one foal. Twins are very rare and, where this occurs, one of the twins would more than likely be very small and would not survive. Horse-donkey hybrids are almost always sterile because horses have 64 chromosomes whereas donkeys have 62, producing offspring with 63 chromosomes. The offspring of a zebra-donkey cross is called a, zebroid. Occasionally zebroids are seen in areas of East Africa where donkeys and zebras graze in close proximity. They are easily identified by stripes on their legs, quarters and, sometimes, their sides. This cross is often found to be very difficult to train as zebra are known for their unfriendly nature and it seems that this trait is passed on to the young.

The offspring of a zebra-donkey cross is called a, zebroid. Occasionally zebroids are seen in areas of East Africa where donkeys and zebras graze in close proximity. They are easily identified by stripes on their legs, quarters and, sometimes, their sides. This cross is often found to be very difficult to train as zebra are known for their unfriendly nature and it seems that this trait is passed on to the young.

Following are the main desired traits for breeding of donkeys (Simoens *et al.*, 1996):

- ❒ Lack of selection because of the difficulty of controlling donkey breeding is, therefore likely wherever a premium is placed on "wild" characteristics of the donkey, such as strength, rather than on docility and productivity for food.
- ❒ The relatively high proportion of males in herds (one male: two females) is another factor that makes control over breeding logistically difficult. The greatest factor contributing to the adaptation of the Ass may be related to its functional anatomy, biology and behaviour.
- ❒ Donkeys are miniature creatures with slender bodies and anatomical built that permit them to support load over 60 per cent of its live weight. They have shown remarkable stamina and resilience to move long hours without food and water.
- ❒ The hooves are the narrowest and smallest of any equid to effect long distant journeys.
- ❒ In equids, the pectinate ligament is more prominent than in ruminants and characterized by sturdy interconnected strands and relatively small intertrabecular spaces

Courtship

Courtship and Access to Semales

In Arid Climates

- ❒ Males control access to resources females need, like water (Moehlman 1998).
- ❒ Able to access females in estrus.
- ❒ Grevy's zebras have similar mating systems.

In Wetter Habitats

- ❐ Donkey stallions have a harem of bonded females that he defends from outside males.
  - ☆ Horses and plains zebras also have this mating system (Moehlman 1998).
- ❐ Feral donkey mares in estrus outside a stallion's territory may attract a number of stallions who fight but may all mate with her. (Klingel 1998).
- ❐ Receptive females stand with hind legs apart and tail held to the side. (Klingel 1998)

Mating Sequence (Klingel 1998)

- ❐ Stallions sniff the female.
- ❐ Female kicks towards him in a ritualized fashion and travels forward.
- ❐ Stallion follows, driving her for an average distance of 20 m (65.6ft).
  - ☆ Tame and feral donkeys and onager males all drive or chase the females a certain distance before mating.
- ❐ When female halts, male mounts, and copulates.

Reproduction

- ❐ Females may be in oestrus for up to 8 days. (Grinder *et al.*, 2006)
- ❐ Females in a group do not all come into oestrus at the same time (Moehlman 1998)
- ❐ Peak breeding seasons vary in feral populations according to timing of wet/dry seasons. (Grinder *et al.*, 2006)

Gestation and Birth

Gestation

- ❐ 1 year

Birth

- ❐ Female gives birth to one or two foals.
- ❐ Weight of foals: 25 kg (55 lb).

Life Stages

Foal

- ❐ Can begin nibbling plants at 5 days when incisor teeth erupt (Moehlman 1998).

- ❒ Regularly browse on vegetation by 2 weeks, but depend on mother's milk for fluid (Moelhman 1998).
- ❒ Weaned at 12-14 months (Grinder *et al.*, 2006).
- ❒ At one year, foal still spends most of its time close to its mother (Grinder *et al.*, 2006).

Adult

- ❒ Sexual maturity for males by 2 years; can produce sperm earlier (Grinder *et al.*, 2006).
- ❒ Females reach sexual maturity at 1.5 yrs but don't usually breed until 2-3 yrs (Grinder *et al.*, 2006).
  - ✰ Observed wild donkeys in U.S. deserts and reported females have first foals later, at age 3.5 to 4 years (Moehlman 2002).

Longevity

- ❒ Averages 20 years in wild (Klingel 1990).
- ❒ 47 years for domestic donkey/burro (Nowak 1999).

Mortality

- ❒ Feral donkeys in U.S. face mountain lion and wolf predators.
- ❒ Feral donkeys in U.S. subject to Wild, Free-Roaming Horse and Burro Act of 1971 (Grinder *et al.*, 2006).
  - ✰ Law requires protection, management, and control of horses and burros on public lands.
  - ✰ National Park Service is exempt from this law and may remove burros dead or alive from its lands.
- ❒ Wild donkeys in Australia are culled by hunters in helicopters and on the ground.

Chapter

5

Feeds and Feeding

Donkeys can endure on very poor diet and nourishing conditions. The nutritional necessities have not been correctly expounded. Some documentation delivers rations for donkeys similar to that of horses. However, a respectable daily ration for donkeys should be contingent on a greater extent on the breed, age, purpose and reproductive status of the Ass and the essential daily food consumption necessary for it to meet both physiologic and physical needs. Ebangi *et. al.* (2004), investigated the nutritional intake of the N'dama and Crossbred donkeys in the Sudano-sahel region of North Cameroon and found substantial differences in feed consumption and weight gain; work force and response to seasonal changes between the N'dama and the cross bred donkey. Generally, the donkey eats in minor amounts over a longer period and can handle 1.5 per cent of its body weight per day in dry matter because of its effectual and anatomically designed digestive tract. The efficacy of the donkey's digestive capabilities over the donkey may also be linked with the functional levels and interplay of its gut enzymes and microbial flora. While a donkey can acquire most of its energy needs from organizational carbohydrates, there is a great controversy on the type of feed sources a donkey may be fed. An archetypal feed ration for a donkey should contain of forage as the main source and other sources of proteins, fat, vitamins and minerals with regular access to fresh drinking water. It is important that the ratio of legumes to grass may not exceed 1 in 4 parts. It is usually prudent to avoid nourishing potatoes or anything from Cabbage family, rhubarb leaves, cereals, onions, leeks, garlic or fermented foods.

Donkeys with access to grass all year round won't require much hay, even during the winter months. You should control the amount of grass and hay the donkeys eat each day to maintain ideal body condition. Grazing should always be considered as supplementary to straw, which should make up the majority of a healthy donkey's diet. Talk to one of our welfare advisers to find out more information about donkey nutrition.

Straw

We recommend feeding donkeys quality barley straw as it is high in fibre and low in sugar, and closely resembles the food that a donkey would eat in the wild. Constant access to straw allows a donkey to eat to appetite without consuming too many calories. Too many additional calories adds a risk that the donkey could put on excess weight, which has associated risks of developing health conditions such as laminitis and hyperlipaemia. Oat straw may be useful for old or underweight donkeys as this usually has a slightly higher nutritional value than barley straw. Wheat straw is very fibrous and has lower energy values, but may be fed to young healthy donkeys with a good set of teeth. Linseed straw is best avoided since the seed is poisonous to donkeys unless it has been boiled, and it is very difficult to ensure that no seed is present in the straw. If straw is in short supply then priority must be given to using it as a feed source and alternative bedding such as shavings should be used.

Hay

If you plan to feed your donkeys hay during the winter months, ensure you have enough forage in store to see you through the winter as supplies could run out before the winter is over. Never feed donkeys mouldy hay as they may have potentially fatal allergies to the mould. It is best to try and build up a good relationship with your hay supplier to make sure you have a consistent supply of good, clean hay suitable for feeding to your donkeys. Different types of hay available include:

- Meadow Hay is a natural mix of grasses made from grass grown on old pasture and is suitable for feeding to donkeys.
- Seed Hay is also good for donkeys. It is a planted crop of specific grasses, such as rye or timothy, which the farmer makes from the stems remaining after the grain has been taken.
- Hay produced from cow pasture usually has higher energy levels and may be less suitable if fed on its own. However it could be fed mixed with a higher ratio of straw.

Ragwort in hay is very poisonous to donkeys and can kill equines. It can be quite hard to distinguish once it has dried; this is why it is important to know and trust your hay supplier.

Weather conditions tend to influence the cutting season - late May to July. Remember late cut hay will have lower energy values, which may suit your donkeys if they are overweight and require forage with a lower energy content, but not if they are elderly/underweight donkeys that need feeding up. If the hay is cut later in the year a lot of the goodness will have gone out of the grass and some of the grass will have gone to seed. This type of hay provides a much lower energy source than early cut hay and for that reason it is fine for feeding to donkeys. If the weather in May is good the farmers might make hay in the first week of June and get a second cut at the end of July. This 'second cut' hay is usually lower in energy value and again is fine for most donkeys.

Freshly cut hay should be stored in a dry barn for at least three months before feeding. Do not suddenly introduce freshly cut hay to the donkey's diet as it could cause colic or laminitis. Reduce the risk by mixing the new hay with the previous year's hay, or mix it with straw over a few days so there is a gradual change over. If hay is in short supply in your area (or if it is very expensive) then you could look at the following alternatives.

Haylage is semi-wilted grass that has been allowed to dry to only 55-65 per cent dry matter (as compared to 85 per cent in hay). The grass is baled, compressed and sealed in tough plastic and the resultant forage is virtually dust-free, highly palatable and nutritious. Once the plastic wrapping is broken (deliberately or accidentally) fungal spores start to grow so the haylage must be used within three to four days (less in warm weather) or discarded. This is why it is better suited to those with more than one or two donkeys to feed. If there are any signs of mould or yeast growth on a bale once opened it should be discarded, as should any uneaten haylage as this could be toxic to the donkeys.

Haylage can be very variable in terms of nutritional levels; some haylage may be too high in energy to feed to donkeys. If you are unsure about the suitability of locally available haylage as donkey food, we would recommend having it analysed (most of the large horse feed companies provide this service for a small fee) or feeding a commercially available equine haylage marketed as 'laminitic safe'.

Silage

Silage is not suitable for feeding to donkeys because the moisture level is usually too high, with a low pH, as well as a low fibre and high protein level.

High Fibre Pellets

There are many brands on the market selling high fibre pellets. Products marketed for equines prone to laminitis are a good choice because they are usually high in fibre and low in sugar. High fibre pellets are a good choice if you need more than grass, hay and straw to build up the weight of an old or underweight donkey. Care must be taken that the donkey does not eat the pellets too quickly, as it may cause colic, so add water and mix with a small quantity of low sugar chaff when introducing pellets for the first time. High fibre pellets can be soaked down to a mash, which is particularly useful for donkeys with poor teeth. Avoid any pellets containing cereals grains as these are not suitable for feeding to donkeys. Products marketed as 'mixes' are usually cereal grain based and again not suitable.

Short Chop Chaff Products

Consider feeding your donkey chaff, a mixture of chopped up hay and/or straw. There are many types of chaff on the market. These contain variable amounts of chopped rye, timothy or alfalfa grasses and oat straw. Some have added oil, molasses, minerals, herbs or hoof growth supplements whilst others are high fibre and molasses free. Chaff products marketed for equines prone to laminitis are good for donkeys that have difficulty eating straw, hay or grass due to poor dentition, and can be used as feed supplements or fed ad lib as a total straw/hay replacer. Always choose a chaff that is 'laminitic safe' and preferably with a sugar content of less than 8 per cent when feeding donkeys.

Dried Sugar Beet Pulp

Sugar beet pulp, a by-product of the sugar industry, is a useful food given in small amounts to tempt a sick donkey to eat or to help an elderly donkey or underweight donkey gain condition. Sugar beet is a source of succulent, nutritious, digestible fibre when added to the feed, although it cannot be fed as a replacement to hay. We recommend unmolassed sugar beet to help prevent laminitis and unnecessary weight gain. Dried sugar beet pulp is available in shredded or cubed form and must be thoroughly soaked before feeding and used within 24 hours once wet. Soaking times vary so refer to the manufacturer's instructions. There are now some quick-soak unmolassed sugar beet products on the market which soak in under ten minutes although it's always advisable to check the product is fully soaked before giving them to your donkeys.

Succulents

Fruits and vegetables can be fed in small amounts (one or two a day) to provide variety and encourage a healthy appetite. They are a worthwhile addition to the normal food ration in winter and early spring when fresh grass is not available. Avoid feeding potatoes, anything from the brassica family, onions, leeks, garlic, stoned fruit and anything which is old, fermented or mouldy as these are toxic to donkeys.

Carrots, apples, bananas, pears, turnips and swedes are all safe and usually very popular with donkeys. Ensure that chopped fruit and vegetables are cut in a way that minimises the risk of choking, such as in sticks.

Minerals and Vitamins

Donkeys on a forage only diet may not obtain all of the required vitamins and minerals from the straw, grass and hay in their diets. Therefore, we recommend that in addition to the standard food you supply that all donkeys will benefit from the addition of a vitamin and mineral supplement in the form of a balancer or unmolassed equine mineralised block, designed to supplement their diet all year round to prevent any deficiencies. Blocks designed for other livestock may be toxic to donkeys as some contain inappropriate mineral levels.

We have successfully used TopSpec Donkey Forage Balancer for donkeys on weight loss or maintenance diets and TopSpec Comprehensive Balancer for those with extra nutritional needs such as growing, breeding, underweight, ill or elderly donkeys.

Water

Water is perhaps the most essential of all nutrients since without it life cannot continue for longer than a few days, or less in adverse conditions. Clean, fresh water should be freely available at all times. Remember to check water supplies in cold weather to ensure they have not frozen.

Some donkeys, particularly the old or unwell, don't like drinking very cold water, so it may be a good idea to provide warm water in the winter, particularly if you are unsure if your donkey is drinking enough. If it is possible provide access to several sources of water to increase the choices available to your donkey, enriching their environment and encouraging them to drink sufficient amounts.

General Considerations

- ✔ All feed stuffs should be of high quality.
- ✔ All equines are sensitive to toxins that can be found in spoiled feeds.
- ✔ All feed stuffs should be free from mould.
- ✔ All changes to their diet should be made gradually, over at least 7-14 days.
- ✔ Donkeys prefer to browse for their bulk and fibre throughout the day.
- ✔ Donkeys prefer to eat little and often. Provide aid-lib barley straw.
- ✔ Do not over feed your donkey; check the body condition of your donkey regularly.
- ✔ Donkeys do not need high levels of sugar in their diets.
- ✔ If in any doubt about the energy value or the quality of any feed, it is advisable to seek expert advice.
- ✔ Always provide a suitable balancer or mineral lick and permanent access to a clean water supply.
- ✔ Never feed grass clippings and ensure that your neighbours are aware of the dangers of doing this as it can lead to colic.

The Energy Requirements for Work for a Donkey

1. Carrying a load over level ground on dirt tracks[1]	
Liveweight of donkey	120 kg
Distance travelled	15 km
Load carried	40 kg
Energy cost of walking (1.40 J/m/kg)	2520 kJ
Energy cost of carrying (2.3 J/m/kg carried)	1380 kJ
Total net energy of work	3900 kJ
Proportion of total energy cost of work used in walking	65 per cent
Total net energy cost of work as a proportion of maintenance requirement	0.31
2. Ploughing a field for 2.6 hours at an average draught force of 730N with a team of four[2]	
Live weight of donkey	120 kg
Distance travelled	5.5 km
Work done per donkey	1004 kJ
Efficiency of pulling	0.25
Load carried	0.25 kg
Energy cost of walking (2.0 J/m/kg)	1380 kJ
Energy cost of doing work pulling	4016 kJ
Total net energy cost of work	5336 kJ
Proportion of total energy cost of work used in walking	25 per cent
Total net energy cost of work as a proportion of maintenance requirement	0.42
3. Carting a load over level ground at an average draught force of 140N on laterite roads[3]	
Liveweight of donkey	120 kg
Distance travelled	15 km
Work done	2100 kJ
Efficiency of pulling	0.35
Energy cost of walking	2520 kJ
Energy cost of doing work pulling	6000 kJ
Total energy cost of work	8520 kJ
Proportion of total energy cost of work used in walking	30 per cent
Total net energy cost of work as a proportion of maintenance requirement	**0.67**

1: Data from Dijkman pers comm. 2: Data from Hagmann and Prasad (1994). 3: Data from Slingerland (1989)

Conclusion

Donkeys are non-selective grazers and can consume a wide variety of feedstuffs. As free-ranging animals, they seldom need any additional feed. However, when the animals are expected to produce extra work or when the range has been heavily depleted, supplementary feeding may be required. A wide variety of crop residues, *e.g.* cereal stovers and leguminous forages and crop by-products such as maize and sorghum bran are suitable feeds for donkey supplementation. The supplementary feeding of donkeys, however, should only be for special purposes and for select groups of animals. Ideally, donkey feeding should not compete with that of other domestic animals whose supplementary feeding may be more beneficial to the farmer.

Dangers of Overfeeding

Besides obesity – which comes with its own set of health risks – overfeeding donkeys can lead to potentially fatal conditions. These include *founder* or *laminitis*, which can cripple the donkey for life if he survives the initial bout. Donkeys are prone to various metabolic diseases, including *insulin resistance*. If your donkey begins sprouting fat rolls, typically on the neck and hips, consult your vet for advice on your animal's diet.

Common Health Problems Associated with an Inappropriate Diet

- ❐ Be aware of the risks of **laminitis** (inflammation of the sensitive tissues in the hoof causing pain and lameness) from excessive consumption of grass or cereal-based feeds (stressed grass, *i.e.*, frosty conditions also increase the risk of laminitis). Donkeys are particularly at risk of this condition when the grass is lush and growing quickly. Fields which have been previously fertilised may pose a greater threat. Laminitis is a serious condition that requires urgent veterinary advice.
- ❐ Limit cereal (grain) based diets. In one study, donkeys that had been fed a diet including cereal-based concentrates were more likely to have gastric ulcers when compared to donkey's fed on a fibre concentrate or forage-only diet.
- ❐ Donkeys may also become **obese** on unrestricted grazing. The sensible use of fencing will enable you to control your donkey's grass intake. Please keep your fence well maintained and check it regularly, moving as appropriate. The rotation of grazed areas can also be very useful in helping to control parasitic worms.
- ❐ Feeding cereal-based concentrates is generally not recommended due to the risk of health problems (unless specifically recommended by a veterinarian).

Chapter 6

Production Systems

Donkey farming may be connected to its resolve, herd size, usage and atmosphere. Effective use of donkeys would depend on many conditions (Kumwenda, 2004). Socio economic and cultural factors play an important role in donkey keeping.

Following are some of the donkey production systems available:

- Extensive systems
- Mixed and Small-scale systems
- Semi-intensive systems
- Intensive systems

Extensive Systems

Under these systems, the donkey however inadequate by space is permitted to graze freely during the course of the day on the field, drinking and resting when and wherever it wants. This is identical with grazing in the wild, whereby feral donkeys may have widespread range coverage depending on season and obtain ability of needed resources and threat from predators and hunters. The extensive systems are very common in tropics, predominantly in harsh conditions.

Mixed and Small-scale Systems

Donkeys are still more or less an add-on livestock for smallholder rural occupiers where they are used as pack animals. Draught and transport are then the main purpose of keeping donkeys, in a free range system, periodic rest and specialized tasking. These systems are popular throughout sub-Saharan Africa (Starkey, 1994).

Semi-intensive Systems

Under this system, the donkey or pack of donkeys are subjected to intervallic grazing and some limited additional feeding as well as spending part of the day on the field and whole night in the stable, fence or barn. This can be seen in farms where donkeys have been assimilated with other livestock. Dead and life (electric) fences are built to form enclosures that may be used to control grazing as a pasture management strategy.

Intensive Systems

Also known as 'zero grazing', intensive system pacts with single or a small number of donkeys in which they may be tethered or stabled and fed on the spot with adequate quantities of hay, straw, vitamins and concentrate rations. This could be seen in dairy herds, feedlots or with the young (Polidori *et al.*, 2009). However, under traditional management systems, tethering is more or a less a way of limitation or incarceration rather than for a specific production purpose.

Production Models

There are numerous farming models that have been developed so far. The production of various donkey products for export has been shown in the figure below followed by the detailed information.

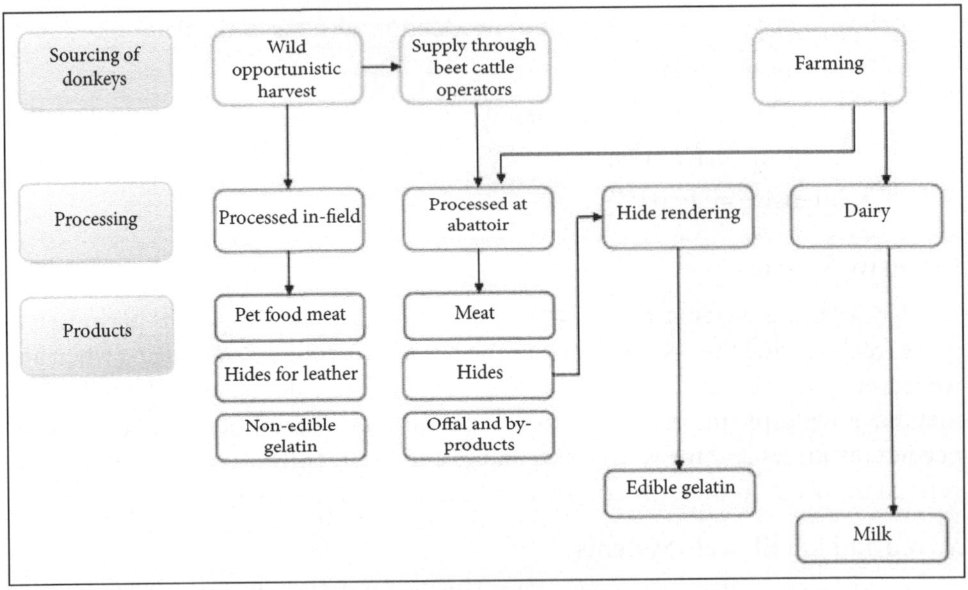

Schematic of Donkey Farming Export Products.

In general, donkeys have been very crucial to pastoral people in arid regions. Nevertheless, donkeys have been always managed less than other livestock (Marshall and Weissbrod, 2011). Insufficient care remains a matter in donkey husbandry.

Donkey husbandry, possession and practice can be useful pointers on the wellbeing and living conditions of many agrarian or peasant economies. Traditional donkey farming consists of a lawn or grazing field, a stable with a drinking point and feeder trough. Nevertheless, most donkeys are usually hitched on the back yard where it may be fed some leftover household remnants.

Traditionally, donkeys may be given limestone or salt as mineral licks, yet care under these circumstances remains much to be desired. A typical donkey farming model should provide facilities for grazing/exercise, shelter, watering adlib, feeding and restraint with back up nutritional supplements to improve performance and output. Wherein, donkeys are kept as a pack lot or for draught purposes such nutritional support is recommended for vitality, repair and energy needs of the animal. Although donkeys tend to express instinctive stubborn behaviour, they can be well tamed and take to organized activities.

Sourcing of Donkeys

For centuries, in rural cultures across the globe, one animal has been an important part of the family, helping to keep farms and villages running.

Not only do millions of people depend on donkeys for practical purposes - many donkeys are seen more and more as smart and loyal pets.

A shocking new demand for a kind of gelatin is killing millions of donkeys every year. The growing trade is known as "ejiao," which refers to the glue that is created from the donkey's skin, also known as "collacoriiasini" or "donkey hide glue". A new report from the Donkey Sanctuary in the UK has demonstrated how extensive this trade is, moving from a centuries-old positive relationship between people and donkeys, both for practical purposes and as smart and loyal pets, to a highly negative one that is killing unthinkable numbers.

Opportunistic harvest (field shooting and feral capture) of feral donkeys provides a short-term resource. However, it is unlikely that this could sustain a profitable industry in the longer term due to the large capital investment required for product processing and the need for consistent, quality-assured supply. Opportunistic harvesting has a number of logistical challenges related to remote and inaccessible terrain and the temperament of feral donkeys. Feral donkeys are difficult to handle and even experienced operators report that this is a challenging undertaking. Harvesting or mustering in the wet season may not be possible due to logistics and the potential for animal welfare issues. Mustering costs are unpredictable.

Currently, the main method used to control feral donkeys is *in situ* aerial shooting, but because the gelatine obtained from donkey skin is destined for human consumption, donkeys intended for this purpose must be killed at an accredited abattoir. As the donkey skin trade is such a lucrative market, other options being considered include live export and farming of donkeys.

Statistics from the report show that the global trade is providing around 1.8 million donkey skins every year. However, more information is emerging from the new trend, and numbers of donkey deaths to satisfy the demand could be between 4 million and 10 million. At the moment, more precise numbers are unclear as the trade is difficult to track and has never before been studied at a large scale.

While destined for different markets, shipments of donkey skins present similar logistical challenges to other contraband, placing their handling and transport to Asia within the arc of wildlife traffickers and other groups accustomed to evading the law. The exports of donkey skins are coming from many places all over the world, including South America and Asia, although the main source is coming from the African market. In Africa, they are often being stolen and crammed together without shelter from the hot sun, whilst being deprived of food and water as they wait to be slaughtered. After their skins are removed the bodies are often burned.

In the long term, breeding programs could be established to select for particular animal attributes (size, meat quality, hide quality). This could be achieved through animal selection, using mares as surrogates and artificial breeding programs. In addition, breeding programs will be high cost and require special yard facilities. It is envisaged that farming will take several (in the order of ten) years to get to a steady state production system. At this point, dairying could be further considered. Culling for biodiversity outcomes and a reduction in grazing pressure would need to be balanced with the opportunity to harvest or capture feral donkeys.

Processing

Meat was originally processed to preserve it, but since the various procedures cause so many changes in texture and flavour it is also a means of adding variety to the diet. Processing also provides scope to mix the less desirable parts of the carcass with lean meat and in addition is a means of extending meat supplies by including other foodstuffs such as cereal in the product.

To comply with Meat Industry Regulations meat and offal slaughtered in-field can only be used for pet food. Regulations specify that for pet food purposes, an animal must be bled immediately after slaughter, butchery of the carcass must be commenced 30 minutes after slaughter and be completed without delay. It is also a requirement that the meat is placed in refrigeration within 4 hours of it being butchered. This would require mobile (vehicle-mounted) refrigeration units to collect and transport meat to a pet food processor.

Meat is a highly perishable products and soon becomes unfit to eat and possibly dangerous to health through microbial growth, chemical change and breakdown by endogenous enzymes.

These processes can be curtailed by reducing the temperature sufficiently to slow down or inhibit the growth of micro-organisms, by heating to destroy organisms and enzymes (cooking, canning), or by removal of water by drying or osmotic control (binding the water with salt or other substances so that it becomes unavailable to the organisms). It is also possible to use chemicals to inhibit growth and, very recently, ionising radiation (however, the last is not allowed in some countries).

Hides removed in the field can only be used for leather production (since gelatin is an edible product and hygiene standards are not satisfied by this method). Hides would need to be folded and transferred to a 'salting camp'. Transfer may not be possible by four wheel drive vehicle due to the inaccessible terrain. A helicopter could be used although the latter is likely be cost prohibitive. Once salted, hides would be palletized and potentially containerized for further transport.

Domestic Consumption

Where donkeys are mustered, they would be contained in yards and drafted to separate those suitable for slaughter. The Meat Industries Act includes feral donkey in the definition of game meat. This means that meat can be processed at a domestic abattoir with a domestic game meat processing license or sold via a game meat slaughter licensee. However, it is considered that there would be a limited market for domestic consumption of donkey meat.

Reproduction Management

In parched circumstances the male or the wild ass stallion factually control access to resources like water from females and therefore, achieve access to the female donkey in estrus. In damper conditions, the donkey stallion controls a harem of bonded females which he defends from other conquering males (Moehlman, 1998). A young foal weighing about 25 kg at birth begins nibbling on plants at 5 days when incisor teeth erupt (Moehlman, 1988), keeping closely with the mother until weaning at 12-14 months of age (Grinder *et al.*, 2006). The age at puberty in the female is 1.5 years, but females may not usually breed until 2-3 years. Sexual maturity in males is attained at 2 years (Grinder *et al.*, 2006). Donkeys have very little sexual dimorphism compared to other members of the equidae family.

Genetically, a donkey has 62 chromosomes as against 64 in the horses. Crosses or hybrids have been registered between a donkey and a mare to produce a mule. Many mules constitute the bulk of the draught power of the equidae family as they have been found to be more resistant to diseases, can carry more loads, trot better and work better than the parents that gave birth to them. A cross between a Jennet (a female donkey) and a Stallion (male horse) is called a hinny. Head and front end of the crosses resemble the male (sire) and the hindquarters look more like the female (dam). Generally speaking, a hybrid mule's head resembles the sire's large donkey head (Clutton-Brock 1981).The mule and hinny are usually sterile off springs (of either sex). The offspring between a male donkey and a female zebra is referred to as a ze-donkey, zebroid or zebrass (American Donkey and Mule Society website, 2009). Artificial Insemination (AI) is not very popular in donkeys. However, with increasing demand for donkey milk and meat, AI and Embryo Transfer applications can become vital tools for genetic selection for increased donkey meat and milk.

Donkey Milk Production System

Donkey milk is such a treasured commodity mainly due to the great care and effort that goes into its production. Milk begins to formulate in the jenny after it has given birth and for two months the little foals are fed with milk directly from their mothers' udders without any disruption from farm workers. Then, the mothers are temporarily separated from their foals and milked by hand every three hours. However, to be willing to give their milk, the mother donkeys must always have visual contact with their foals.

The storage volume of milk in a jenny's body is very small, so it must be milked regularly to collect a significant amount of milk, usually amounting to 1.5 litres per donkey per day. After the last milking of the day, the foals, which are fed by bottle throughout the day, return to their mothers for natural feeding. This process is repeated until the foal is 11 months old. All milking on the farm is done by hand, primarily because the use of electric milking machines would be traumatic to the donkeys. And in any case, the small amount of milk collected from these animals would make electric milking machines obsolete.

After each milking, the pure donkey milk is filtered without the use of chemical processing. Analyses carried out in collaboration with the «Centre d 'Expertise de la qualite des Facultes de Gembloux» (The Quality Control Centre of Gembloux Academic Faculties) has certified the bacteriological stability of milk. This means that the donkey milk that is produced on the farm can be drunk without pasteurisation. The milk is then poured into 200 ml containers and frozen immediately in a minimum temperature of –18°C.

It is thereby preserved for consumption as a sweet and pleasant drink or for the manufacture of high-quality cosmetics. The farm unit works according to European Union hygiene regulations and recommendations and is licensed by the Belgian Ministry of Health.

Chapter 7

Preventive Health Care

Like any other working creature, a donkey that does not feel well will not do good work. The health of a donkey can best be checked by regular, daily contact and familiarity with the donkey when it is healthy. Then, when changes occur, they will quickly be seen and appropriate measures can be taken.

An owner's main task is to prevent any harmful effects of work on the donkeys. Regular, work because it is physical exercise, strengthens the muscles of the animal, improves its circulation and breathing and increases appetite.

On the other hand, overwork – such as pulling or carrying loads that are too heavy and for too many hours in the day without proper time for resting and eating – can be harmful.

Overworked donkeys will suffer from fatigue, loss of weight and general weakening which makes them susceptible to disease.

Respiratory Diseases

Donkeys are asymptomatic carriers of the lungworm *Dictyocaulus arnfieldi*, which can cause respiratory disease/coughing in horses and ponies. It is therefore important to treat both donkeys and in contact horses with ivermectin or moxidectin when they are co-grazed. To prevent respiratory disease it is important to keep them in a clean well ventilated environment, vaccinate against flu, worm them regularly, and be familiar with signs of respiratory disease (coughing, increased respiratory noise/effort/rate, and nasal discharge, appearing unwell).

Castration (Gelding)

This is an operation that really must be performed on most male donkeys. It is usually not done before the donkey is about 10 months old. Because at that time the testicles have not yet descended, it will be necessary to cut them out. This creates an open wound that should drain and may take a couple of weeks to heal, so must be carefully managed to prevent infection. It is most unwise to perform the operation in the rainy season when bacterial and fungus infections are common, but even in the dry season the wound must be kept clean and disinfected twice daily until it closes.

From about the time a donkey is 2 years old it is possible to use a burdizzo which does not create an open wound if it is used over a clean cloth and so is less inviting of infection.

However, its results are not so reliable and it has to be very professionally used.

Also if the donkey is too mature it may retain the unwanted behaviour of a stallion even after it has been successfully castrated with the burdizzo.

Feet

Feet are a vital part of a donkey's working equipment because a donkey which cannot walk, cannot work.

The under-surfaces of each foot should be checked out daily during the rainy season, especially if ground conditions are sticky. They should be regularly cleaned with a hard sharp stick, to make sure no decayed matter stays lodged in them and that no stones or thorns remain to irritate them either. During the dry season, checking once a week will be enough.

In wet conditions donkey hooves may get soft or overgrown, so the growth and condition of the hooves should be watched to ensure that the shape remains as it should be. If necessary, a hoof can be filed to the right shape, this may also be necessary in very young donkeys to correct the shape of the legs. The hoof and the frog can also be shaped straight with a very sharp knife.

In dry conditions, especially where surfaces are hard, donkeys' hooves may crack and split.

Regular rubbing with oil but particularly neat's-foot oil will help to prevent this and a nick made of the hoof at the top of the crack at right angles to it will help to stop the crack spreading.

Hoof knives and hoof-picks are obtainable in shops where horse equipment is sold but are also easily made by a local blacksmith.

If a donkey is lame and the hoof is hot to the touch something is causing inflammation. Look for a wound or a thorn. Very often the inflammation can be reduced by a hot soak (a rag soaked in hot salt water pressed to the inflammation until the temperature equalises) or by soaking the foot in a solution of copper sulphate - blue crystals cheaply obtained from chemists or farmer suppliers. A plastic gallon bottle with the top cut off just enough to insert the hoof is the ideal thing for doing this.

A donkey seldom likes its feet to be wet but one way of keeping a foot standing in a bucket of water is to tie one of the donkey's front feet up with a belt.

Skin

A donkey easily bleeds and its skin is easily broken, so skin wounds (lesions) should be prevented as much as possible by using the correct equipment for work. This should ensure that anything that comes in contact with donkey's skin is soft and does not rub or cut.

If prevention fails and a wound occurs it must be quickly dealt with. First of all it must be kept clean with twice daily washings. Hydrogen peroxide which is fairly cheaply obtainable in chemists is very thorough as an initial cleaner especially if decayed matter is in the wound.

The wound should not be left wet but encouraged to dry as quickly and cleanly as possible. If it is to be covered with oil or Vaseline (or insect-repellent 'Stockholm tar') it must be thoroughly cleaned and dried first.

The application of a strong solution of common salt (sodium chloride) may sting the donkey but it promotes healing while it disinfects. The covering is better if the salt is mixed into a paste using clean water and very well-sifted wood ash. This is particularly suitable for wounds facing downwards such as castration wounds.

Mild alkalis such as cooking soda (sodium bicarbonate) or acids such as vinegar (acetic acid) diluted in water have a similar effect mainly discouraging fungus and bacteria. Brown sugar (not diluted) or honey can help if applied to a wound which has become septic, *i.e.*, inflamed and exuding pus.

Eyes

Often, donkeys suffer from weepy eyes. This is most likely a bacterial inflammation spread by flies. If it is not soon remedied it can get worse and eventually cause blindness for the donkey which would reduce the working effectiveness of the animal.

Washing out the eyes with a very mild antiseptic or even with clean water is usually enough to remedy this condition if caught early enough. Washing can easily be done with a plastic squeezy bottle such as those in which washing up liquid or toilet cleaner is sold. Before such a bottle is used for a donkey's eyes, its original contents must be very thoroughly cleaned out.

Effective mild antiseptics that can be dissolved in water are cooking soda, which is a mild alkali, and common salt.

The washing should be done twice a day. If the eyes do not improve within a week and completely recover within two weeks, then will be the time to use a commercial antibiotic eye powder or spray.

Colic in the Donkey

Colic refers to abdominal pain, a clinical sign with many causes. It is like to the condition in horses but the distribution of common causes and signs can be slightly dissimilar. Numerous types of colic may present inversely, and they are often measured on the basis of history, pain, heart rate, respiratory rate, gut sounds and other clinical exam conclusions. Signs of dehydration are not always obvious as donkeys incline to survive better with dehydration than horses due to their adaptation to parched climates. The heart rate of a donkey with colic is often increased above their average normal of about 44 beats per minute. Their normal respiratory rate is 16-20 breaths per minute and this may also increase with colic, depending on the cause.

Common Causes and Treatment of Colic in Donkeys

- Impactions (59 per cent): mostly pelvic flexure, occasionally small colon
- Unexplained (13 per cent)
- Other (12 per cent): Colitis, peritonitis, stomach ulcers, ovarian disease, fractures, *etc.*
- Spasmodic colic (6 per cent), neoplasia (4 per cent), diarrhoea (3 per cent), pancreatitis (3 per cent)

Treatment

Treating colic in donkeys is very similar to horses, and the goals are to eliminate pain, correct dehydration and restore gut motility and passage of faeces (manure).

Painkillers, oral/IV fluids, periods of starvation and turning out may help.

Surgery

If a diagnosis is not made after a clinical exam, various factors are taken into consideration in deciding if a donkey needs surgery. Factors to consider include the duration of colic, distended intestines palpated on rectal examination (may not always be possible in smaller donkeys), abnormal peritoneal fluid, reduced gut sounds, a large quantity of gastric reflux, perceived severity and unrelenting pain.

Diseases and Pathologies of Donkeys

Disease/Pathology	*Signs*	*Preventive Measure*	*Possible Treatment*	*Outcome*
INJURIES –resulting in fractures/lameness/wounds	Pain, bleeding wound or sore, Lameness, recumbent Weakness, inability to movebroken bone, swelling or inflamed skin or muscle	Proper training Obtain right equipment Inspect and care for tools-greasing, *etc.* Apply equipment correctly Ensure safe working environment	Rest and provide first aid-wash wound, minimize bleeding, relief pain and seek veterinary attention For septic wounds debride, clean and seek professional advice from a Vet If fracture-support with a bandage and seek help from a local veterinarian	
SADDLE SORES **Ulcerating wounds on the back associated with friction an improper padding caused by harness**	Ulcerating wounds or abscesses on the back	Clean saddle and sundry regularly	Rest, clean sores and apply penicillin powder or ointment until healingSeek veterinary advice	Treatment usually rewarding if started early and beast stabled rest during treatment period
LAMINITIS or Founder – affecting the lamina of the hoof. A fatal condition caused by starch overload or grass high in fructan or feeding excessive grain diet	Painful and swollen hoof May discharge fluid Cannot bear weight on the affected limb	Avoid feeding heavy grain diet or sugary foods Recommended commercial high fibre cubes	Rest, clean hoof seek veterinary intervention	Early intervention is desired Dressing and or padding of affected hoof
OBESITY- a metabolic condition associated with overfeeding and over weight	Overweight, gets lazy	Avoid lush pastures and concentrate feeds Daily exercise Monitor weight regularly Increase straw intake and low calorie diet	Correct feeding regime desired	Early intervention good

Disease/Pathology	*Signs*	*Preventive Measure*	*Possible Treatment*	*Outcome*
HYPERLIPEMIA- A fatal condition characterized by depletion of fatty acid stores and the circulation of free fatty acids to the liver and kidney that can cause irreversible liver, kidney and heart damage	Lactating jennies at higher risk than males, Signs are nonpecific including- off feed, fever, dullness, lethargy, rapid weight loss, head pressing, circling and ataxia, seizures and death Predisposing factors include heavy worm egg burden, dental problems, colic, choke, obesity	Vigilance and precaution on feeding Measure of circulating fat (triglycerides)	Hospitalization, nursing care and urgent veterinary intervention may include glucose iv therapy in combination with other support treatment.	Successful outcome depends on early intervention and extent of organ damage
Parasitic diseases-lice, worms, midges, mites, flies and ticks infestations **A donkey can serve as an intermediate host for tapeworms, *Echinococcus granulosum* from cats and dogs defecating in donkey pastures**	Weight loss, ruffled hair coat, skin biting/nibbling. Dull, anorexia, restlessness, itching, irritation, anemia if chronic	Clean, fly proof stable. Avoid wet and dusty stable Regular grooming, Routine dipping	Specific acaricide applications and worming schedules Fly and tick repellents Integrated Pest control	Early intervention usually rewarding
OSTEOCHONDROSIS **Growth disorder associated with imbalanced ratio of Calcium and phosphorus in bones. In case of contracted tendons damage can be permanent if not properly handled**	Fragile bone tissue that can break easily Retarded growth	Proper feeding regime	Proper feeding regime Surgical correction	Early intervention is crucial

Parasites

External Parasites (ectoparasites)

The donkey's habits of mutual grooming, rubbing itself against trees and rolling in sand and ash indicate that it is sensitive to itches, but these habits also rid the donkey of most of the important external parasites.

Parts of the donkey which may harbour parasites are under the tail, between the legs and inside the ears. Parasites in these places can often be removed by hand (but make sure they are properly killed) or by use of tick grease which is commercially available. A non-irritant insecticide such as 'Malathion' mixed with Vaseline is also effective particularly against the mange mite.

Only if absolutely necessary, and if the level of the dip allows them to keep their heads above water (because they cannot swim). Donkeys can be dipped like cattle but some of the dip chemicals in use are harmful to donkeys such as 'Triatix'. Dipping should therefore only be used as a last resort.

Donkeys generally love to be scratched by people especially around and inside the ears. If done as a daily habit, this routine is also an effective way of parasite control.

Internal Parasites (endoparasites)

Especially on poor and over-used pastures, donkeys can be very seriously affected by internal parasites causing lung infections, diarrhoea, weight loss infertility, abortions and eventually an early death. These parasites, therefore, represent a serious economic threat.

Removing manure, rotation of pastures and ensuring that supplementary feed is always clean, are things that can help to keep parasitism under control particularly where donkey numbers are low.

In many countries a twice-annual anthelmintic programme (to control internal parasites) for donkeys, is recommended. Donkeys should be treated for parasites at the end of the rain and again at the end of the dry season. Foals as young as a month old can be included in such a programme.

Chapter

8

Useful Products

There are certain donkey by-products useful to human mankind and have been mentioned below in details:

Donkey Milk

Donkey's milk has a lot of health benefits. Their milk was, of course, hailed as an elixir by the ancients and favoured by Cleopatra for daily bathing but there is now renewed interest in it after Pope Francis revealed he was fed such milk as a baby.

Compared to other types of animal milk, the donkey variety is actually closest to human breast milk, based on its pH level and nutritional makeup. It also contains less total fat than cow's milk, and packs more anti-inflammatory omega-3 fatty acids. Natural substances in donkey milk have been shown to enhance immunity, which may be helpful for people with conditions like asthma, eczema, or psoriasis.

Donkey milk is low in fat with high polyun saturated fatty acids (PUFAs) and resultant advantageous atherogenic and thrombogenic indices. Eicosapentanoic acid and arachidonic acid are present in very small amounts. The high PUFA content and

Do You Know?

Cleopatra reportedly owed her brilliant complexion to her stable of 700 donkeys, who regularly produced enough milk for her to take baths in it. The Ancient Greeks used Donkey Milk for all sorts of ailments; from liver problems to infectious disease.

other beneficial nutrients in donkey milk suggest that it may be useful for human nutrition especially for infants and the elderly.

Following are some of the significant benefits of Donkey Milk:

- Donkey milk is full of protein which helps in rebuilding the muscles. Drinking a glass of donkey milk for breakfast daily helps in muscle building and also reduces body ache caused by intense exercises by retaining fluids and electrolytes.
- Having a glass of donkey milk at night can help your mind calm down if you had a busy and stressful day as it relaxes your exhausted body and eases stressed nerves and fatigue by increasing the energy level. Vitamin B_{12} present in the donkey milk is great for red blood cells (RBC) and nerve tissues.
- Donkey milk helps in normalizing the cholesterol production and works as an anti-acid fluid. It is a good source of vitamin A and B, so it's beneficial for your eyes as well. It boosts the immune system and makes the eye sight normal.
- Various studies suggest that the risk of getting several types of deadly diseases like cancer and diabetes can be reduced if donkey milk is consumed daily.
- Donkey milk contains potassium and phosphorus in abundance which makes it perfect drink for regulating the blood pressure. Phosphorus also makes bones and teeth stronger. Due to the presence of niacin, donkey milk results in better metabolism, reduces acidity and heart-burns.
- Donkey milk contains high amount of anti-oxidants and omega-3 fatty acids which help fight ageing and other chronic ailments.
- If you drink donkey milk regularly, you can reduce the chances of Premenstrual Syndrome. Moreover, donkey milk has been known as a very efficient fertility drink since ages because it lowers ovulation related issues in women by increasing their fertility.

Interesting Fact

While it may seem odd to many of us, the reality is people around the world are taking advantage of this milk and its benefits. For example, in 2011 the BBC reported that more than 50 per cent of the donkey milk produced at a farm outside Bologna, Italy, is sold to pediatric units, for children who can't consume cow's milk. Not to mention, it likely tastes better than it sounds; many people say it's odorless and the texture resembles low-fat cow's milk.

Donkey Milk Benefits for Skin

Donkey milk soap as Natural cleanser

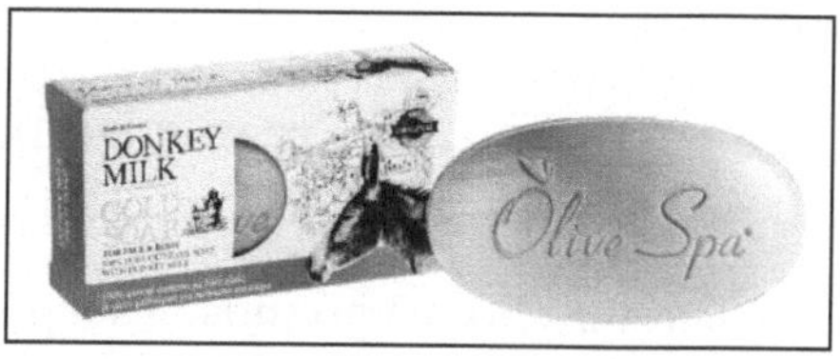

Donkey milk is an amazing natural skin cleanser, as it removes all the dirt and dead-cells from your skin; blackheads or whiteheads becomes softer and can be easily removed by steaming. Just apply some donkey milk or donkey milk soap on your skin and rub gently. Wash it off with lukewarm water.

Donkey milk soap as Natural moisturizer

Organic donkey milk is an amazing natural moisturizer which can help you get rid of lifeless dull and dry skin in winter season.

Donkey milk contains AHA, a compound mostly found in several expensive cosmetics and anti-aging products. An exfoliation with donkey milk, honey donkey milk or oatmeal scrub donkey milk soap can do wonders on your skin. Use the soap with all these ingredients and gently scrub your face and neck with it for about 15 minutes and then wash your face with fresh water.

Donkey milk can also make your skin feel silky and rejuvenated if your skin is dull and lifeless. Donkey milk cleanses the pores and moisturizes the dry skin.

Donkey milk soap as Natural toner

If you are looking to buy a natural and chemical-free toner, then you should try donkey milk soap first. Mix few drops of donkey milk soap and green tea together in a bowl. Dab a cotton ball in it and apply it all over your face. Leave it on for a few minutes, and wash it off. You can clearly see an instant glow and radiance in your skin if you do this regularly.

Donkey milk soap can heal sun burnt skin

Donkey milk in donkey milk soap can help you soothe and heal the sun-burnt skin patches. Apply donkey milk soap daily on your skin to become fairer.

Donkey milk soap benefits for hair

Donkey milk proteins in donkey milk soap can help you get shinier and healthier hair. Donkey milk contains casein and whey proteins which are great for healthy hair growth. Donkey Milk proteins play an important role in making hair thicker and healthier.

Calcium boosts hair growth and prevents hair loss, proteins and lipids make hair roots stronger; whereas vitamins A, B_6, biotin and potassium make your hair shiny, healthy and beautiful.

Other than protein, keratin is also an important element for hair growth; calcium is necessary in order to create keratin, and donkey milk is full of calcium. Applying daily donkey milk soap to your body may help dull and dry hair become healthier and thicker.

Creamy Donkey Milk Soap Recipe

This soap is made with donkey milk for an extra creamy feeling on the skin.

Serves: About 3 pounds of soap

Ingredients

- ❒ 9 Bar Birchwood Mould
- ❒ Silicone Liner for 9 Bar mould
- ❒ 5.3 Apricot Kernel Oil (15 per cent)
- ❒ 8.8 oz. Coconut Oil (25 per cent)
- ❒ 11.5 oz. Olive Oil Pomade (33 per cent)
- ❒ 2.5 oz. Cocoa Butter Cubes (7 per cent)
- ❒ 7 oz. Palm Oil (20 per cent)
- ❒ 4.9 oz. Sodium Hydroxide Lye
- ❒ 11.6 oz. Donkey Milk
- ❒ 2 oz. Oatmeal Milk and Honey Fragrance Oil
- ❒ Titanium Dioxide

Instructions

Milk Prep

When lye is added to milk, it heats up to around 200°F. This scorches the milk, causing discoloration and an unpleasant odour. To prevent scorching, it's important keep temperatures as cool as possible. Our preferred technique is to freeze the donkey milk and add the sodium hydroxide lye flakes directly to the frozen milk. To keep the milk extra cold, you can place the container into a larger container with ice. In this project we skipped this step, but it's a personal preference. Measure out 11.6 ounces of donkey milk and pour it into ice cube trays (in this case we used Medium 9 Ball Silicone Moulds). Allow the milk to fully freeze. Then, place a container appropriate for mixing lye on a scale and weigh out the frozen cubes. Although it was measured previously, some milk might be lost in the freezing/transferring process. If you're missing any, pour cold donkey milk into the container until you have 11.6 ounces. You can also use cold distilled water.

Safety First

Suit up for safe handling practices! That means goggles, gloves and long sleeves. Make sure kids, pets, and other distractions and tripping hazards are out of the house or don't have access to your soaping space. Always soap in a well-ventilated area.

Lye Prep

With safety gear on, slowly add about ¼ of the lye flakes directly to the frozen donkey milk. Use a stainless steel spoon to stir the lye flakes and donkey milk together. Slowly, the flakes will begin to melt the donkey milk cubes. After a few minutes of stirring, add another ¼ of the flakes and continue stirring for several minutes. Continue this process until all the lye flakes have been added to the donkey milk. Continue stirring and stirring. The donkey milk will continue to melt as more lye is added. Once the milk has fully melted, continue to stir to make sure all the lye flakes have completely dissolved. Because the temperatures are low, the lye flakes do not dissolve as quickly. Listen and look for lye flakes on the bottom of your container; lye flakes can take a lot longer to dissolve in cooler temperature liquids. Patience is key! Any undissolved lye in your soap could cause skin irritation or even lye burns, so take extra time to make sure all the lye is dissolved. Once all the lye has been added, the milk may become yellow. Below, our milk reached about 90°F and developed a very slight yellow colour. This colour of milk is suitable for this recipe. But, if you'd like to make sure your soap batter is as light as possible for your colour palette, placing the container of frozen milk into an ice bath helps keep temperatures cooler. If you'd like a harder bar of soap that releases faster from the mould, you can add sodium lactate to the milk solution. Use 1 teaspoon of sodium lactate per pound of oils in the recipe. For this recipe, you'd add 2 teaspoons sodium lactate.

Colour Prep

To ensure that the titanium dioxide blends smoothly into the soap batter, we recommend micronizing it before dispersing it in oil. Please note this is an optional tip but it does help with the titanium dioxide clumping in the soap. To micronize colorant, simply use a coffee grinder to blend the colorant to break up any clumps of colour and prevent streaks of white from showing in the final soap. We like to use a coffee grinder that has a removable, stainless steel mixing area for easy cleaning. Disperse 3 teaspoons of titanium dioxide into 2 tablespoons of lightweight liquid oil, like sweet almond oil or sunflower oil. In this recipe, we used a slightly more concentrated colorant-oil ratio to avoid adding too much extra oil.

Fragrance Prep

Measure 2 ounces of oatmeal milk and honey fragrance oil into a glass, fragrance oil safe container and set aside.

1. Completely melt and combine the coconut oil, olive oil pomade, cocoa butter, apricot kernel oil, and palm oil. Once the oils have cooled to 130

degrees or below, add the Oatmeal Milk and Honey Fragrance Oil directly to the oils. Usually fragrance oils are added at trace, we know this fragrance behaves so well and thin trace is not a concern, it can be added first.

2. Add the milk and lye solution slowly to the oils. As you add the milk, begin pulsing your stick blender. Because the milk and lye solution is on the cooler side and the recipe contains hard oils that solidify at cooler temperatures, adding the lye and milk solution gradually can help prevent false trace.
3. Once all the lye solution has been added, continue to stick blend until you reach thin to medium trace.
4. Add all the dispersed titanium dioxide, and use the stick blender to stir and blend the colorant into the batter using short bursts.
5. Once the colorant is completely incorporated, pour the soap into the mould. Tap it firmly on the counter to eliminate air bubbles, and use a spoon or spatula to spread the soap evenly into the mould.
6. Using a small stainless steel spoon, create texture on the top of the soap.
7. Once you're happy with the top, insert the dividers into the mould. Push the dividers all the way to the bottom of the mould. Spray the top of the soap with 99 per cent isopropyl alcohol to help avoid soda ash. Milk soap has a tendency to heat up due to the sugar in the milk. To keep it cool, place the soap in the fridge or freezer for at least 3 hours, up to overnight. Then, allow it to sit in the mould for about 3-4 days. Soap that is placed in the fridge or freezer for the first few hours of saponification can take a little longer to unmould. If you find the silicone liner is not pulling away from the sides of the mould easily, give it another day to unmould. Remove the soap from the dividers and allow the bars to cure for 4-6 weeks.

In addition to alimentary benefits, consumption of donkey milk has been allied with enhanced immune responses. Consumption of donkey milk increases interleukins and tumor necrosis factor as compared to the levels seen with consumption of donkey milk. These findings suggest that donkey milk might be examined as an advantageous dietary component for immune compromised individuals as well as the sick and elderly. Donkey milk contains other factors that may be beneficial for human health. It has been suggested as an alternative to human and cow milk for individuals with allergic diseases, inflammatory disorders and atopy. Donkey milk has been recommended for feeding to infants who are allergic to cow milk.

In an experimental study, some infants were provided with cow milk with the addition of soy protein with poor results. When donkey milk with added triglycerides was provided to infants, the milk was well-tolerated by all patients. Patients receiving donkey milk had no recognizable clinical reactions and gained more weight.

Donkey Meat

Tough or tender, one thing we know for sure is that the use of donkey meat isn't new. In China, dating back to the Ming Dynasty in the early 15th century, donkey meat was introduced as a cheaper substitute for horse meat. At around the same time in the south of France, donkey meat was also experiencing a boom. The tradition there actually continued until somewhat recently. One French sausage producer explains that they don't use donkey meat in their product anymore because donkey slaughtering is now restricted. Their original donkey-and-pork saucisson d'Arles is now made with pork and beef.

Donkey meat occasionally enters the human food chain via adulteration. However, donkey meat has potential as a beneficial alternative to traditional red meats because of its high nutritional profile and components. Donkeys represent a potential threat to human health via meat infected with *T. gondii* and *Neospora*. The study by Gharsa *et. al.*, included in EVJ's online collection on Working Equids, also highlights the importance of donkeys as a potential source of bacterial species carrying virulence factors, which may have important public health implications.

Donkey meat is the culinary term for the flesh of donkey which is used as an ingredient in many dishes. In countries where donkeys are still used for work or milk production (largely North Africa and Middle East), they are slaughtered at an older age when they have reached the end of their useful working life. This may account for the reputation that donkey meat has gained for being unacceptably tough and suitable for salted meat-based products (Polidori *et al.*, 2008).

- ❒ Donkeys have potential as a source of tissues and molecules that can be of use in human medicine: human cardiac disorders are an important cause of morbidity and mortality and heart valve disorders may be treated by prosthetic valve replacement.
- ❒ Donkey pericardium is a potential alternative bioprosthetic heart valve material because it is thin, has minimal calcification values and possesses high tensile strength.
- ❒ Donkeys produce a variety of peptides that may be beneficial to man.
- ❒ Donkeys are now commonly used for the commercial production of antibodies.
- ❒ Peptide HP-6, a peptide derived from donkey serum albumin, may: promote the proliferation of cells related to the haematopoietic system; enhance mouse haemopoiesis function; and increase resistance to chemotherapeutic injury.
- ❒ Donkeys are also used in some parts of the world at high altitudes for the production of anti-venom drugs.

- LH pentapeptide, derived from donkey serum, inhibited the growth of transplanted tumour cells and stopped their infiltration into other organs.
- In addition to their potential contributions to the advancement of human health, donkeys may also provide mental and moral support to human individuals.
- The human–animal bond that develops with donkeys can help individuals with motivation and development of psycho affective and psycho cognitive processes.
- Rehabilitation sessions with donkeys can help in identification of individual strengths so that motivation may be strengthened.

A research by students from Egerton University's Animal Science Department confirms that donkey meat is healthy for consumption. The Turkana and Chinese say the meat has a unique taste, texture and health benefits. A 100 gram of donkey meat has fat, proteins, potassium, phosphorus and sodium - making it the most nutritious meat in the world. It is also said that donkey meat has five times less amount of total fat which reduces chances of getting obesity and diabetes.

Donkey Skin

Donkeys have potential as a source of tissues and molecules that can be of use in human medicine: human cardiac disorders are an important cause of morbidity and mortality and heart valve disorders may be treated by prosthetic valve replacement. Donkey pericardium is a potential alternative bioprosthetic heart valve material because it is thin, has minimal calcification values and possesses high tensile strength. Donkeys produce a variety of peptides that may be beneficial to man. Donkeys are now commonly used for the commercial production of antibodies. Peptide HP-6, a peptide derived from donkey serum albumin, may promote the proliferation of cells related to the hematopoietic system; enhance mouse hemopoiesis function; and increase resistance to chemotherapeutic injury. This study in rats showed promising effects *in vivo* and *in vitro* and additional human studies may be warranted. LH pent peptide, derived from donkey serum, inhibited the growth of transplanted tumour cells and stopped their infiltration into other organs. Donkeys are also used in some parts of the world at high altitudes for the production of anti-venom drugs. Horses, commonly used for this purpose in other parts of the globe, are not well-adapted to high altitudes. Future research may be appropriate to determine whether this type of anti-venom production may be commercially appropriate.

The donkey skins are being used to make gelatine for a product called ejiao, a product that has been used as a traditional Chinese medicine for thousands of years. Ejiao (said: eh-gee-yow) is a hard gel that can be dissolved in hot water or alcohol to be used in food or drink, or in beauty products such as face creams. Believed to improve blood circulation, ejiao is used as a blood tonic by people with anaemia, low blood cell counts or reproductive problems. Ejiao, or donkey skin gelatine, is considered one of the three treasures of traditional Chinese medicine. It is used to treat a wide range of ailments from simple colds to insomnia and impotence, and demand in the Chinese market is soaring like never before. Millions of donkeys are slaughtered all around the world and their hides transported to China to be melted into the miracle gelatine that many believe will keep them looking youthful and even prolong their life.

"Ejiao is a medicine with ancient roots and has been promoted as a product worthy of emperors," the report says, explaining that traditional herbalists in China claim that ejiao can increase libido, slow aging and prevent disease. But ejiao has not been recognized as having medicinal properties by western medicine. The sweet-tasting ejiao has proven to be an effective tonic/medicine for women.

It provides the following health benefits:

1. Invigorates blood circulation, moistens the lungs and treats yin deficiency (by reinforcing body fluids and nourishing the blood) and cures dizziness, insomnia, coughing and palpitation, which result from lack of blood or yin;
2. Helps women who suffer from such ailments as coughing up blood and/or passing blood to stop bleeding;
3. Curing irregular menstruation, stabilizes the fetus (if the woman is pregnant) and prevents miscarriage. As ejiao invigorates blood circulation, it can treat lack of blood and blood stasis, both of which are major causes of irregular menstruation. In addition, ejiao contains such elements as lysine, arginine, histidine, cysteine, calcium and sulfur, which help prevent miscarriage;
4. Helps prevent and treat cancer. Ejiao can be used to prevent cancer, and it can be used as a supplementary medicine for patients who receive radiotherapy and chemotherapy; and
5. Improves physical appearance. As ejiao invigorates blood circulation, women who eat it often tend to have rosy cheeks and glossy skin. In addition, ejiao helps strengthen bones and muscles and improves the quality of sleep.

Market Demand for Donkey Products

Donkeys are the backbone of many farmers in developing countries. But if current trends continue, the world's rural poor may soon need to find a new beast of burden. The animals' ranks have thinned dramatically in many African countries: Kenya's donkey population, for example, has fallen by half to 900,000 since 2009. The primary cause is neither disease nor declining demand for live donkeys, but instead a burgeoning market for their pelts.

In recent years, the market in ejiao—a product made from boiled-down donkey skins mixed with herbs and other ingredients—has grown massively in China, putting millions of donkeys at risk of slaughter or poaching.

In South Africa, poor farmers who rely on donkeys as beasts of burden and modes of transport have reported having their animals stolen, only to later find their skinless carcasses. South Africa's Society for the Prevention of Cruelty to Animals has confiscated more than 1,000 donkey hides on route to China in the past year. The growing trade in donkey skins could signal the extinction of the animal in South Africa. "In two years, there won't be any donkeys in South Africa".

Kenya has the highest number of donkey slaughterhouses in Africa. According to an animal welfare researcher, who has asked for anonymity to protect his research and sources, many African countries have expressed alarm at the opening of donkey slaughterhouses in their villages. Residents in Ethiopia burned down the first slaughterhouse before it was even operational. The demand for donkey hide is outstripping supply. Around 1.8 m donkey hides are traded per year, while global demand is estimated between 4-10 m skins, states the report. This has raised the price of donkeys in some countries, making them unaffordable for many people who use them to take goods to market, cultivate land, and fetch water. The cost of a donkey in Burkina Faso, for example, has increased from £60 (₹5346) in 2014 to £108 (₹9367) in 2016. Activities to procure donkey skins are largely unregulated and through the animal welfare partnerships in more than 40 countries, agencies have seen evidence of welfare issues for donkeys at every step, from sourcing to transport and finally to slaughter. The donkeys are frequently sourced from donkey-dependent communities, particularly in Africa and South America, and the trade has impacted on their livelihoods and even economic stability.

But there is also a move towards donkey farming at scale, with a strong focus in China. Again, this industry to produce meat, milk and skins is largely unregulated and with likely growing welfare concerns.

The consumption of donkey meat is a popular tradition in some parts of China, where it is highly expensive. Some groups in Africa eat donkey meat, considered an inexpensive and accessible source of animal protein, but only from the slaughter of donkeys too old or incapacitated to work. In many communities, its consumption is considered taboo.

In the last two decades, China's donkey population — once the world's largest — contracted from 11 million to less than 6 million, while per capita GDP increased almost tenfold, from just over \$1,800 in 1995 to more than \$14,000 in 2015. Rapid industrialization and urban migration greatly reduced the need for donkeys in rural areas, while rising affluence in cities boosted demand for consumer products like ejiao. The result has been a surge in skins imports from developing countries, often with scant regard for the suffering of donkeys or the lives of people who depend on them.

According to UK based charity, The Donkey Sanctuary, there are currently 15 countries who have stood against the trade in DHG. They state that some have gone through the formal legislative route like Mali, Niger, and Burkina Faso, whereas others have blocked export by preventing certain things like veterinary inspections and transport permits from being issued. Fourteen of those are African whilst the other country to make a stand is Pakistan.

Chapter 9

Worldwide Market Niche of Donkey Milk Farms

Donkey milk farms have flourished all over the world. Italy, Bosnia and Herzegovina, Serbia, Switzerland, Turkey and France in Europe, and Chile or Mexico in South America, are just examples of countries in which donkey milk is reaching limits which were at first improbable.

Donkey milk is very similar in composition to breast milk and is widely used as a treatment for allergies such as lactose intolerance and eczema in babies and young children. Donkey milk is similar to mare milk and human breast milk in that it is relatively poor in protein and fat but rich in lactose. The casein to whey protein ratio is intermediate between human breast milk and cow milk. Gross composition of milk differs by the mother's lactation stage, with ash and protein content showing a declining trend, but pH, percentage of whey protein, and amino acid content remaining the same.

Donkey milk keeps the skin hydrated, smooth and youthful and products contain between 5 per cent (soap) and 50 per cent (face cream) donkey milk. The combination with other natural ingredients and essential oils perfectly complements the properties of the milk. All these cosmetics are biologically and ecologically friendly and are made by professional hands. Soaps are made using the method of hot saponification by a soap-maker who learnt his craft from an old traditional master in Marseille. Soap is formed and put into its boxes by hand. The laboratory for the cosmetic products is near Dijon and donkey milk sweets are made in Vichy by the same company that produce the famous 'bonbons de Vichy'. If you are travelling around the south of France during the summer you should not hesitate to go and visit the farm and their donkeys in the mountains (Commune de Couflens – chemin des Cascades de Leziou).

A female ass gives between 0.2 to 0.3 litres of milk a day. The milk of a donkey was disapproved by the U.S.D.A until 2001. It is frowned upon in the average American society. A donkey dairy farm may range from a few to over 600 animals. It seems that we are getting back to ass milk to raise children in the earliest infancy, notably when the child is of delicate health (as it happens in India). Ass milk has not been quite totally abandoned, but if 25 or 30 years ago, a few well looked-after asses were easily found in the city to provide milk nourishing young babies, it is no longer the case today.

Although the market potential is huge donkey farms normally start selling only pure milk with ordinary people who have bronchial problems as the main consumers—and then it soon moves on to cosmetic production which normally starts with homemade soap production. One litre of donkey milk sells between 35 and 55 euros in European countries due to its rarity and limited supply. Soap sells for 7 euros per bar.

Donkey milk production is still a stranger for some countries, not everybody likes the idea of having a donkey farm in the neighbourhood or drinking this special beverage. Alen Jusupoviæ, a 23-year-old student of agricultural engineering in Sarajevo, was intrigued by the idea of starting a donkey farm after learning that a similar business was thriving in neighbouring countries. In 2012, he set out to have the first in Bosnia and Herzegovina.

Turkey's first donkey farm in the northwestern province of Kýrklareli is the first also aiming at producing donkey milk for consumption. The project started in 2009, and the farm is now in use with a stock of nearly 180 donkeys, but the farm is always on the lookout for more of them, as the owner complains of not being able to find donkeys due to an insufficient number of them in the country. The production of milk is yet to start in the farm due to formalities between Usta and the relevant authorities. Usta said he had completed the necessary paperwork and was now waiting for a response.

Export of Meat

Donkey is one of the most important domestic species that in many countries contributed to economic development because of its rusticity and its ability to survive and produce meat and milk also in hard environmental conditions (Polidori *et al.*, 2008). These species, in the past, were particularly used in large farms in ploughing the fields or other works, particularly in North Africa and in the middle east of Europe, where they were also an important means of transport. Nowadays, the higher donkey head number is concentrated in the tropical and arid regions of the world where infrastructures are not well developed, where they are still used as working animals.

Dubai and Pakistan are expected to earn millions of dollars annually in foreign exchange by exporting donkeys, said a senior official at the livestock department in Khyber Pakhtunkhwa province. Pakistan has the world's third largest population of donkeys with more than 5 million animals, while China stands at number one. In a bid to boost the country's exports to earn much needed foreign exchange, the Live Stock Department in Khyber Pakhtunkhwa has announced that it will develop donkey farms — the first of its kind in the country. Pakistan is planning to sell donkeys to China after an investment of ₹1 billion in its 'Donkey Development Programme', in a bid to attract Chinese investors in the north-western Khyber-Pakhtunkhwa province.

Steps are being initiated to increase the population of donkeys in the province as the animals are highly valued in China for its hide which is used to manufacture medicines among other things. To facilitate export of meat, skins and gelatine some markets (including China) require that export accreditation and oversight of the establishment is administered by the Australian Government (rather than NTG) and this is referred to as a Tier 2 facility. A Tier 2 facility has strict controls for meeting food hygiene, environmental and animal welfare standards.

A protocol between Australia and China for the export of donkey meat and edible hides does technically exist. However, it has not been operationalised by government to government agreement on health certification which defines processing, refrigeration, packaging and transport requirements. This means that even where donkey products are processed in a Tier 2 facility the protocol does not enable export. There are no specific protocols for the live export of donkeys to any international market. As the same genus as horses, donkeys would be considered as an equivalent commodity. Importing country requirements for horses to China have been agreed. However there has been some challenge with the demonstration of meeting these requirements over a number of years. For other countries where import protocols have not been negotiated it would be possible to organize an import permit. Horses are not considered 'livestock' under the Exporter Supply Chain Assurance System. It would be particularly important to demonstrate that animal welfare issues were fully considered and addressed to maintain a 'social license' for live export.

Increasing reports of mixing donkey meat with beef for commercial purposes have been reported in some countries. The work of Yman *et al.* (1995) in differentiating horse meat from donkeys and their hybrids by electrophoretic separation of albumin can be useful indicators in detecting tainted meat and meat products from the horses, donkeys, mules and hinnies. Staining of the starch gel for carboxylesterase activity permits differentiation of most horses from donkeys while mules and hinnies cannot be distinguished from horses by their esterase activity alone.

Donkey Meat Farms

China and some countries in Africa are known for their conventional donkey meat based cuisine. Yet, their production systems are quite different. If we thought about any other kind of meat producing farms, such as the ones that we have to produce chicken meat or beef, we could easily change the species bred there by donkeys and we would realize what a donkey meat farm is like. These farms are commonly placed in South American and Eastern European countries (like Ukraine, Peru or Mexico) where, though there it is not legal to eat donkey meat, or better said, to slaughter donkeys with alimentary purposes, it is possible to produce donkeys and export them to countries such as Canada or China, where they can be slaughtered and transformed into traditional dishes. Animal number per pen ranges from 50 to 100 animals and depending on the country, they are likely to produce only jennies, jack stocks or both sexes.

Africa would be better adjusted to the example of a backyard production farm, in which a family or a whole tribe keeps their donkey herd among other animals such as cattle or donkeys. Animals are kept in groups which range from 30 to more than 300 animals, generally looked after by women or young men. These kinds of farm are also the main producers of donkey hides.

The analysis was embarked on using the Breedcow software package. It is available from the Queensland Department of Agriculture and Fisheries website (www.daf.qld.gov.au/animal-industries/beef/breedcow-and-dynama-software).

A number of setups were developed to address the inter-related issues of export requirements, potential products and farming systems. These issues are briefly described below.

Export Requirements

The demand for donkey hides is mostly driven by the Chinese market. Export of meat and hides to China entails dispensation in a Tier 2 facility which has stringent controls for meeting food hygiene, environmental and animal welfare standards.

To test the effect of the cost of transport on the gross margin a scenario has been included where donkeys are purchased at the property (farm gate) and the cost of transport is considered further along the value chain.

Hides could be processed in a Tier 1 abattoir and sold into an alternative export market. A Tier 1 abattoir could be serviced in a regional centre such as Katherine. The cost of transport of donkeys from the station to Katherine (600 km) has been included in the analysis where appropriate.

Donkey Meat and By-products

When donkeys are processed in a Tier 1 facility meat and by-products could be consumed domestically. However there is currently only a very limited domestic market for donkey meat or by-products in Australia.

When donkeys are processed in a Tier 2 facility meat and by-products (brain, eyelashes, heart, hooves, intestine, kidney, liver, penis, tails, tendons, tongue and tripe) could be exported to China.

To test the effect of income from meat on by-products on the gross margin a range of scenarios have been used – export to China; sale for pet meat; and disposal as waste at no cost.

Reaching Steady State Farming

The modelling commenced, here, undertakes a stable state farming system which will entail 5-10 years to achieve. Even, where a traditional beef production property is used, there might be additional capital costs to alter it to a donkey enterprise or miscellaneous species operation. For example, supplementary or diverse configurations of fencing and yard systems may be required. A steady state farmed system would primarily rely on capture of feral donkeys. Anecdotal proof suggests that these donkeys can be especially challenging to handle. There are also particular animal welfare deliberations relating to transport and overheating. These interactive and physiological characteristics are likely to result in additional (as yet un-quantified) costs compared with beef production to the primary producer in order to reach a steady state farming system.

Carrying Capacity

Feral donkeys have been detected to eat grasses to ground level then uproot them, promoting soil erosion and land degradation. This could lead to a greater impact on the native pasture resource for a given level of pasture consumption than for cattle. This grazing behaviour has been assumed to be linked with forage scarcity and the effect could be managed through pasture consumption tactics.

Scenario	*Product*	*Market*	*Processing*	*Notes*
1.	Hides	China	Darwin	
2.	Hides	China	Darwin	No transport cost included (*i.e.* buyer pays transport costs)
3.	Hides	China	Darwin	
4.	Hides, meat, by-products	China	Darwin	
5.	Hides, meat, by-products	China	Peterborough	
6.	Mammoth hides, meat, by-products	China	Darwin	
7.	Beef	International	Darwin	

The following was calculated for each scenario:

1. **Gross margin** – the total sales revenue to the enterprise minus the cost of stock sold, on a per animal basis.
2. **Total gross margin** – the total sales revenue to the enterprise minus the cost of stock sold, on an annual basis.

Donkey and beef cattle enterprises have been analysed on a Gross Margin per Adult Equivalent (AE) basis to allow comparison.

The gross margin analysis for donkey farming done by the Northern territory indicates that if access to the Chinese market were available the farming of donkeys for hides, meat and by-products in the NT could in the long term be commercially viable. However, this conclusion needs to be considered in the context that reaching a steady state farming system may require high capital costs and a long period (5-10 years) of low return on investment at the enterprise level. A farming enterprise is also dependent on access to a Tier 2 abattoir in South Australia (existing) or an NT that can process donkeys.

It also needs to be reinforced that there is currently no operational protocol for market access to China.

Main Sources of Donkey Food

Class	*Sources of Nutrient*	*Recommendations*
Protein	Forages, hay, haylage Protein pellets or cubes	Ideal during winter. Must be properly stored and guarded against moulds, ragworts and pests
Carbohydrate	Freshly cut hay, haylage	Highly palatable but must be fed in very small amounts.
Fibre	Barley straw Short chopped up chaff (SCUC) Dried sugar beet pulp (DSBP)	SCUC is useful in donkeys with poor dentition useful as a total hay/straw replacer declared as 'laminitic safe' DSBP should not be molassed base to avoid risk of laminitis (follow manufacturer's instructions)
Minerals/Vitamins	Straw, grass and hay Salt granules or blocks	Free access to Salt blocks or granules meant for donkeys such as equine mineralized blocks are recommended
Water	Clean fresh water	Free access to fresh clean water. Avoid serving frozen water or very cold water. Provide several sources of drinking points on a grazing field to encourage feeding and grazing ad lib.
Succulents and Treats	Fruits and vegetables Carrots, bananas, pears, apples, turnips and Swedes Mixture of oats, oat bran and corn	

Glossary

Asiatic wild ass: The related **Asiatic wild ass**, sometimes called the Asian wild ass or the half-ass (*E. hemionus*), is usually known by the local names of its various races.

Asino: An asino is the Italian word for "donkey".

Ass: An ass is either a male or female donkey.

Background: A position that attracts little attention.

Bray: Donkeys make a loud sound developed in order to maintain contact with other **donkeys** over wide spaces in the desert. This is called a bray. While **donkeys** in captivity have no need to stay in contact with other **donkeys** over space, they will still bray for a variety of reasons.

Breeder: A person who keeps and takes care of animals in order to produce more animals of a particular kind.

Brush: To clean or smooth something (such as hair) with a brush.

Burro: A burro is the Spanish word for "donkey".

Canine Teeth: Nearly all male donkeys and some female donkeys have canine teeth. These erupt at about 5 years, so if they are present, they also tell us that the donkey is at least 5.

Colds and Flu: Donkeys can get infections similar to the common cold in people. The signs of such an infection include: Increased temperature (39°C +), Dullness, Lack of interest in food, Discharge from the nose – either watery or yellowy or whitish, and Coughing.

Colt: A colt is a young male donkey which is less than four years of age.

Diet: Donkeys typically like to eat grass, though they also eat shrubs and desert plants. Instead of ripping the grass right away with their teeth, donkeys grab the plant with their lips, pull it in their mouth, then rip it with their flat teeth and grind it down for swallowing.

Donkey hybrids: A male donkey (jack) can be crossed with a female horse to produce a mule. A male horse can be crossed with a female donkey (jenny) to produce a hinny.

Donkey shed: The name of a donkey house is usually a "donkey shed." It can be called "donkey house," as well. Donkeys can also be kept in corrals.

Donkey: The **donkey** or ass (*Equus africanus asinus*) is a domesticated member of the horse family

Ecology: The common donkey or *equus asinus* is a very durable animal and can be found almost anywhere in the world. The first donkey that was not domestic was originally found in hilly grassy areas or in northern Africa, where it was used as a pack animal and sometimes even food. Even though the domestic donkey has been able to adapt to almost any environment from snowy regions to extremely hot regions, it will always prefer the warmer regions and if left to become feral, will end up returning to a area with a warmer climate.

Equus asinus asinus: It is the scientific name for the donkey based on the principle of priority used for scientific names of animals.

Filly: A filly is a young female donkey which is less than four years of age.

Foal: A foal is a baby male or female donkey up to one year old.

Gelding: A castrated male donkey.

Halter: If a donkey's head is restrained, it can be led or held for procedures such as injections. A halter can be made from a piece of sisal or cotton rope. Avoid using nylon rope against the skin. A simple slip halter can be made with loops.

Head collar: Some animals learn to slip halters off over the ears. A head collar is better. Head collars are suitable for donkeys, but should have buckles to adjust the size of the straps around the nose. This way the head collar can be made big enough to go around a donkey's nose.

Hinny: A hinny is the result of breeding between a female donkey and a male horse.

Indian wild ass: The Indian wild ass (*Equus hemionus khur*) also called the Ghudkhur, Khur or Indian onager in the local Gujarati language, is a subspecies of the onager native to Southern Asia.

Jack: A jack is a term for a male donkey.

Jenny: A jenny (or jennet) is a term for a female donkey.

Lice: Both biting (*Damalinia* spp) and sucking (*Haematopinus* spp) occur mainly at the mane but can cover the whole body.

Mare: A female donkey.

Moke: A moke is a British term for a donkey.

Molly: A molly is a term for a female mule.

Mule: A mule is the result of breeding between a male donkey and a female horse.

Neoplastic Skin Disorders: Sarcoids are a special type of growth in the skin that affects only horses and donkeys. They can appear as a single or multiple red lumps on the skin or lumps under the skin. They can look very different, and are found in different places. They can also occur in wounds. Sarcoids can be difficult to deal with as removal doesn't always cure the problem, if it is possible at all. Sometimes, they can get worse if attempts are made to ligate or surgically remove them.

Pet Donkeys: Donkeys are lovable, intelligent animals that aren't necessarily all workers. Some are kept as pets. They may live in a stable at night and a pasture during the day to graze; others may live in a pasture day and night. It is important that donkeys have access to some type of shelter for protection from wind and rain. A sturdy lean-to will do. Donkeys originate from warm climates; in a region with ice and snow, donkeys need a warm stable or barn.

Rescued Donkeys: Hundreds of donkeys require rescuing each year. Some are abandoned, others abused and seized. Some donkeys are surrendered to rescues by owners unable to care for them any longer. In the United States, some are feral donkeys captured by the Bureau of Land Management and placed for adoption to control protected herd populations. These rescued donkeys live in animal sanctuaries operated by nonprofit animal rescue groups. The size and scope of these sanctuaries vary greatly from a few donkeys to hundreds. Understanding the needs of donkeys, rescue groups provide shelter and pasture acreage for grazing. Sturdy fences made of wood planks or wood posts and wire enclose the pastures and keep the donkeys safe.

Rig: A rig is an entire male donkey with no signs of external testicles.

Ringworm: This is a disease caused by parasites that live on the skin – mainly the hair. It makes the hair weak so that it falls off easily. It usually affects young donkeys or weak ones with other problems. Humans can get infected if they get into contact with donkeys with ringworm.

Sarcoptic mange: Skin condition characterized by alopecia, intense pruritis, scab formation and ulcerations.

Stallion: A stallion is a male donkey that has not been gelded (castrated).

Strangles: This is a disease that looks like a cold, but after a few days pus may be seen in the nostrils and the glands in the neck may become swollen. These glands will burst in time. The donkey may cough, have a high temperature, be generally dull and will probably be off its food or have difficulty eating.

Straw: We recommend feeding donkeys quality barley straw as it is high in fibre and low in sugar, and closely resembles the food that a donkey would eat in the wild.

Stress – *n.* a state of mental tension and worry caused by problems in your life, work or other things.

Temperature: The normal temperature of a donkey is between: 36.2- 37.8°C.

Threat of Extinction: The wild donkey has been captured for centuries for breeding and to use as pack animals, which is causing the species to dwindle to the hundreds in the wild which is dangerously close to extinction. The donkey is also hunted for sport, food, and medicine in villages and some other hunting trips.

Ticks: Common tick species identified in donkeys are *Amblyoma variegatum*, A. lepidum, *A. gemma, A. choarens, Boophilus dicloratus, B. annulutus, Hyalomma truncatum, H. marginatum, H. anatolicum, Rhipicephalus pulchellus, R. evarsi* and *R. turanicus.*

Trauma: A very unpleasant experience that causes someone to have mental or emotional problems usually for a long time.

Veterinary: Relating to the medical care and treatment of animals.

Whipworm: Whipworms cause irritation of the anal region making the animal restless and causing it to rub its tail against a wall or post. Infected animals do not feed properly and can lose condition.

Wild Donkeys: The wild African ass, from which domestic donkeys originated, survives in the countries making up the Horn of Africa, according to Ronald M. Nowak, editor of "Walker's Mammals of the World." Wild donkeys living elsewhere in the world are not actually wild donkeys. They are feral domestic donkeys, often called burros.

Working Donkeys: Donkeys have served as pack and draft animals for about 5,000 years. Today, millions of people living in developing countries rely on donkeys to till fields, transport produce and other products to market, carry water, pull carts and work in mills. Such donkeys live wherever is most convenient for their owners. Some live on farms in corrals, and some live in urban environments — even inside houses. Donkeys work in developed countries, too. They are highly affectionate and patient. For this reason, donkeys serve in equine therapy programs for disabled riders. Donkeys sometimes serve as foal or stable companions because of their calming effect on horses. These donkeys typically live in stables and spend time out in pastures during the day.

Worms: Donkeys have a number of parasitic worms in the gut which may sometimes be found in the dung. Maggots of the bot fly live in the stomach and are also found in the dung.

Yearling: A yearling is a young male or female donkey between one and two years of age.

Zebra: African animal that looks like a horse and has black and white stripes covering its body.

References

Aganga A A and Tsopito C M 1998. A note on the feeding behaviour of domestic donkeys: a Botswana case study. *Applied Animal Behaviour Science* **60:** 235 - 239.

Aganga AA, Aganga AO, Thema T and Obocheleng KO. 2003. Carcass Analysis and Meat Composition of the Donkey. *Pakistan Journal of Nutrition* 2 (**3**)**:** 138-147, 2003. Accessed online at www.pjbs.org/pjnonline/fin113.pdf.

Al-Sadi, B.I., Singh, A.P., Al-Badrany, M.S. 1990. A study of animal neoplasms in Iraq. *Indian Journal of Animal Sciences* 60 (**12**)**:** 1419–1424.

American Donkey and Mule Society 2009. URL: www.lovelongears.com/about_mules.html.

Benoit, A. 1985. L'Ane, dressage, élevage. pp 29–30 in L'âne: aspects historiques, culturels, zootechniques. Ethnozootechnie 37. Societé d'Ethnozootechnie.

Berry, C. and Kokas, J-A. 1989. Donkey Business II. A Guide for Raising, Training and Showing Donkeys. Off and Running Print Management.

Blench, R., Jode, A., Gherzi, E. 1998. Donkeys in Nigeria: History, Distribution and Productivity.

Blench, R., Macdonald, K. 2000. A history of donkeys, wild asses, and mules in Africa. **In** Blench and MacDonald (Eds.) The Origins and Development of African Livestock: Archaeology, Genetics, Linguistics and Ethnography. pp: 339-355.

Bougler, J., Duplan, J-M., Zongo, D. 2005. Caractéristiques et répartition mondiale des espèces animales exploitées par l'Homme. *In* Thewis *et al.*, Eds. Manuel de Zootechnie comparee, pp: 45-68.

Boyt, W.P., Mackenzie, P.K.I., Pilson, R.D., Leavis, H. 1972. The importance of the donkey (*Equus asinus*) as a source of food and a reservoir of trypanosomes for Glossina mortisans Westw. *Rhodesia Science News* 1972: 19–20.

Browning, D.G., Scheifele, P.M 2001 Vocalization of *Equus asinus*: the hees and haws of donkey brays. *Acoustical Society of America Journal* 115**(5):** 2485-2485.

Chen, S., Xu, L., Liu, Y., Li, Q., Wang, D., Wang, X. and Liu, T. 2013. Donkey pericardium as an alternative bioprosthetic heart valve material. Artif. *Organs* **37,** 248-255. 21.

Clutton-Brock, J. 1981. Domesticated Animals from early times. University of Texas Press, Austin and British Museum (Natural History), London.

Clutton-Brock, J. 1999. A Natural History of Domesticated Mammals. Cambridge University Press.

Coppock, D.L., Ellis, J.E., Swift, D.M. 1988. Seasonal patterns of activity, travel and water intake for livestock in south Turkana, Kenya. *Journal of Arid Environments* 14 **(3):** 319-331.

CPhl 2009. Accessed online www.cphi.com/europe/home.

Cuddeford, D., Pearson, R.A., Archibald, R.F., Muirhead, R.H. 1995. Digestibility and gastro-intestinal transit time of diets containing different proportions of alfalfa and oat straw given to Thoroughbreds, Shetland ponies, Highland ponies and donkeys. *Animal Science* **61:** 407–417.

De Rose, P., Cannas, E. and Reinger Cantiello, P. 2011. Donkey-assisted rehabilitation program for children: a pilot study. *Ann. Ist Super. Sanita* **47,** 391-396.

Department of Agriculture and Food, Western Australia. 2013. Feral donkey. Accessed online at www.agric.wa.gov.au/pest-mammals/feral-donkey.

Department of Sustainability, Environment, Water Population and Communities, Commonwealth Government. 2011. Feral horse (*Equus caballus*) and Feral donkey (*Equus asinus*). Fact Sheet. Accessed online at www.environment.gov.au/system/files/resources/b32a088c-cd31-4b24-8a7c-70e1880508b5/files/feral-horse.pdf.

Dijkman J T 1992. A note on the influence of negative gradients on the energy expenditure of donkeys walking, carrying and pulling loads. *Animal Production* **54:** 153 - 156.

Fahmy, S.W. 2004. The health and husbandry of donkeys used by Zabbalin rubbish collectors in Cairo, Egypt in Fielding and Starkey eds Donkeys, people and development: A resource book for the Animal Traction Network for Eastern and Southern Africa. CTA, Wageningen, Netherlands. pp. 238-240.

Fernandez, G.P., Segura, A., Herrera, M., Velasco, W., Solano, G., Gutierrez, J.M. and Leon, G. (2010). Neutralization of *Bothrops mattogrossensis* snake venom from

Bolivia: experimental evaluation of llama and donkey antivenoms produced by caprylic acid precipitation. *Toxicon* **55,** 642-645.

Feseha, G.A., Mohammed, A., Yilma, J. M. 1991. Vermicular endoparasitism in donkeys of Debre-Zeit, Ethiopia: strategic treatment with Ivermectin and Fenbendazole. pp 156–166 *in*: Fielding and Pearson (eds), Donkeys, mules and horses in tropical agricultural development. *Proceedings of colloquium held 3–6 September 1990, Edinburgh, Scotland.* Centre for Tropical Veterinary Medicine, University of Edinburgh, UK.

Fielding D and Krause P 1998. Donkeys. The Tropical Agriculturalist. *CTA*. Macmillan.

Fielding, D. 1986. Tropical Equine Project: Preliminary equine survey in Africa. *Draught Animal News* **6:** 18-19.

Fielding, D. 1987. Donkey power in African rural transport. *FAO World Animal Review* **63:** 23–30.

Fielding, D., Pearson, R.A. (eds). 1990. Donkeys, mules and horses in tropical agricultural development. *Proceedings of colloquium held 3–6th September 1990, Edinburgh, Scotland.* Centre for Tropical Veterinary Medicine, University of Edinburgh, UK.

Forhead, A. J., Smart, D., Smith, R.F., Dobson, H. 1994. Endocrine and metabolic responses to transportation in periparturient donkeys. *Animal Reproduction Science* 36(**3-4**)**:** 315–327.

Forrest, A.R.R., Carnegie, P.R. 1994. Identification of gourmet meat using FINS (forensically informative nucleotide sequencing). *Biotechniques* 17(**1**)**:** 24-26.

Grinder, M., Krausman,P., Hoffmann, R. 2006. *Equus asinus. Mammalian Species* **794:** 1-9.

Groves, C. P. 1986. The Taxonomy, Distribution and Adaptations of Recent Equids. pp. 11–51 **In:** Meadow R H and Uerpmann H P (eds) Equids in the Ancient World. Wiesbaden: Ludwig Reichert.

Hagmann J and Prasad V L 1994. The use of donkeys and their draught performance in smallholder farming in Zimbabwe. Project Research Report II. Conservation Tillage for sustainable crop production systems. GTZ/AGRITEX/IAE/P.O. 415, Borrowdale, Harare, Zimbabwe.

Hardman P. 2012. Donkey cheese, anyone. Articles in the UK Daily Mail 15 December 2012. Access online at www.dailymail.co.uk/news/article-2248431/Donkey-cheese-At-1-000-kilo-expensive-world-But-I-getting-milk-donkey-make-ass-anyone.html#ixzz3yUyWQGef.

Izraely, H., Chosniak, I., Stevens, C.E., Shkolnik, A. 1989. The donkey: coping with low quality feed. *Asian Journal of Animal Science* **2:** 289–291.

Johnstone, J.C. 2004. A Biometric Study of Equids in the Roman World. Ph.D Thesis, Department of Archaeology, University of York. 742p.

Jones P 1997. Donkeys for Development. H and W. Printers, Louis Trichardt, Northern Province. South Africa.

Kefen, E., Beja-Pereira, A., Han, J.L., Haile, A., Mohammed, Y.K. and Dessie, T. 2011. Eco-geographical structuring and morphological diversities in Ethiopian donkey populations. *Livestock Science* 141(**2-3**): 232-241.

Kimura, B., Marshall, F. B., Chen, S., Rosenbom, S., Moehlman, P. D., Tuross, N., Sabin, R. C. Peters, J., Barich, B., Yohannes, H., Kebede, F., Teclai, R., Beja-Pereira, A., Mulligan, C.J. 2010. Ancient DNA provides insights into African wild ass phylogeny and donkey domestication. *Proceedings of the Royal Society B: Biological Sciences* 278(**1702**): 50–57, doi: 10.1098/rspb.2010.0708.

Kingdon, J. 1997. The Kingdon field guide to African mammals. Academic Press, San Diego, 476p.

Klingel, H. 1998. Observations on social organization and behavior of African and Asiatic Wild Asses (*Equus africanus* and *Equus hemionus*). *Applied Animal Behavior Science* **60:** 103-113.

Kugler, W., Grunenfelder, H-P., Broxham, E., Monitoring Institute. 2008. Donkey Breeds in Europe. Inventory, Description, Need for Action, Conservation. Report 2007/2008. Monitoring Institute for Rare Breeds and Seeds in Europe and Save Foundation. 62p.

Kumwenda, W.F., 2004. Factors affecting the efficient use of donkeys in Malawi in Fielding and Starkey eds Donkeys, people and development: A resource book for the Animal Traction Network for Eastern and Southern Africa. CTA, Wageningen, Netherlands. pp. 148-151.

Li, H., Yan, S., Wang, Z., Shen, X. and Rao, P. 2011. *In vivo* and *in vitro* effect of peptide HP-6 derived from donkey serum albumin on hematopoietic system. *Zhongguo Zhong Yao Za Zhi* **36,** 2259-2264. 22.

Li, H., Zhong, J., Huang, M., Li, Z. and Rao, P. 2013. Antitumour effects of pentapeptide derived from donkey serum albumin both *in vitro* and *in vivo. Life Sci.* **92,** 883-889.

Lopez, A., Chavira, H., Granillo, A. 1994. Foot and hoof disorders in the underpriviledged equids of Mexico. pp. 57–64 *in*: Bakkoury M and Prentis R A (eds). Working Equines. Proceedings of the second international colloquium held 20–22 April 1994, Rabat, Morocco. Actes Éditions, Institut Agronomique et Vétérinaire Hassan II, BP 6202-Instituts, 10 101 Rabat, Morocco.

Maloiy G M O 1973. The effect of dehydration and heat stress on intake and digestion of food in the Somali donkey. *Environ. Physio. Biochem.* **3:** 36 - 39.

Maloiy G. M. O. and Clemens E T 1980. Gastrointestinal osmolality, electrolyte and organic acid composition in five species of East African herbivorous mammals. *J. Anim. Sci.* **51:** 917 - 924.

Maloiy, G.M.O. 1970. Water economy of the Somali donkey. *American Journal of Physiology* **219:** 1522–1527.

Marino R, Albenzio M, della Malva A, Muscio, A and Sevi A. 2015. Nutritional properties and consumer evaluation of donkey bresaola and salami: Comparison with conventional products. Meat Science. Volume 101, March 2015, Pages 19–24. Accessed online at http: //www.sciencedirect.com/science/article/pii/S030917401400480X.

Marshall, F., and Weissbrod, L. 2011. Domestication Processes and Morphological Change: Through the Lens of the Donkey and African Pastoralism. Current Anthropology, Vol. 52, No. S4, the Origins of Agriculture: New Data, New Ideas (October 2011), pp. S397-S413. URL: http: //www.jstor.org/stable/10.1086/658389. Accessed: 17/06/2013 22: 02.

Marshall, K., and Ali, Z. 2004. Gender issues in donkey use in rural Ethiopia in Fielding and Starkey eds Donkeys, people and development: A resource book for the Animal Traction Network for Eastern and Southern Africa. CTA, Wageningen, Netherlands. pp. 62-68.

Mattioli, R.C., Zinsstag, J., Pfister, K. 1994. Frequency of trypanosomiasis and gastrointestinal parasites in draught donkeys in The Gambia in relation to animal husbandry. *Tropical Animal Health and Production* 26(**2**)**:** 102–108.

McDonnell, S.M. 2002. Equid play ethogram. *Applied animal behavior science* **78:** 263-290.

Meutchieye F, Kwalar N N and Nyock RAF. 2013. Donkey Husbandry and Production Systems. *In*: Navas Gonzalez, FJ. (ed.) Current Functionality and Donkey Production. Accessed online at www.researchgate.net/publication/271212674_Donkey_Husbandry_and_Production_Systems.

Moehlman, P.D. 1998. Feral asses (*Equus africanus*): intraspecific variation in social organization in arid and mesic habitats. *Applied animal behaviour science* 60: 171-195.

Mofya, R. 2004. Social consequences of introducing donkeys into Zambia in Fielding and Starkey eds Donkeys, people and development: A resource book for the Animal Traction Network for Easten and Southern Africa. CTA, Wageningen, Netherlands. pp. 140-142.

Mrema, M. 2004. Economic and gender issues of donkey use in Kweneng and Kgatleng Districts, Botswana *in* Fielding and Starkey eds Donkeys, people and development: A resource book for the Animal Traction Network for Eastern and Southern Africa. CTA, Wageningen, Netherlands. pp. 166-171.

Mueller PJ, Protos P, Houpt KA and Van Soest PJ 1998. Chewing behaviour in the domestic donkey (*Equus asinus*) fed fibrous forage. *Applied Animal Behaviour Science* **60:** 241 - 251.

Mutua J. 2004. Some challenges to the use of donkeys in Kenya in Fielding and Starkey eds Donkeys, people and development: A resource book for the Animal Traction Network for Eastern and Southern Africa. CTA, Wageningen, Netherlands. pp. 99-101.

Orhan, Y., Boztepe, S., Erturul, M. 2012. The domesticated donkey: 1 – Species characteristics. *Can. J. App. Sci.* 2012; 4(**2**): 339-353.

Orlando, L., Metcalf, J.L., Alberdi, M.T, Telles-Antunes, M., Bonjean, D, Otte,M., Martin, F., Eisenmann, V., Mashkouri, M., Morello, F., Prado, J.L., Salas-Gismondi, R., Shockey, B.J., Wrinn, P.J., Vasil'ev, S.K., Ovodov, N.D., Cherry, M.I., Hopwood, B., Maleb,D., Austin,J.J., Hännia, C., and Cooper, A. 2009. Revising the recent evolutionary history of equids using ancient DNA. PNAS, 22 December 2009, vol. 106, no. 51.

Oudman, L. 2002. Donkeys for traction and tillage. Agrodok Series 35. CTA eds, 84p.

Pandey, V. S., Khallaayoune, K., Ouhelli, H., Dakkak, A. 1994. Parasites of donkeys in Africa. pp. 35–44 in Bakkoury M and Prentis R A (eds). Working Equines. Proceedings of the second international colloquium held 20–22 April 1994, Rabat, Morocco. Actes Éditions, Institut Agronomique et Vétérinaire Hassan II, BP 6202-Instituts, 10 101 Rabat, Morocco.

Parsonson I M. 1998. The Australian Ark. A history of domesticated animals in Australia. CSIRO, Victoria. Accessed online at books.google.com.au.

Pearson R A 1994. Energy requirements of working equids. *In*: Final Report: Digestion et nutrition de l'âne. Contract TSZ-0268F (EDB), INRA, Dijon.

Pearson R A, Nengomasha E and Krecek R C 1995. The challenges in using donkeys for work in Africa. Paper presented at the ATNESA Workshop "Meeting the Challenges of Animal Traction" 4 - 8 December 1995. Ngong Hills, Kenya.

Polidori, P., Beghelli, D., Mariani, P., Vincenzetti, S. 2009. Donkey milk production: state of the art. *Ital.J.Anim.Sci.* vol. 8 (Suppl. 2), 677-683.

Polidoria P, Vincenzettib S, Cavalluccia C and Beghelli D. 2008. Quality of donkey meat and carcass characteristics. Meat Science 80 (2008) 1222–1224. Access online at https: //www.researchgate.net/publication/51777935_Quality_of_donkey_meat_and_carcass_characteristics.

Reid, S.W.J., Fowler, J.N., Smith, K.T. 1991. Detection of papilliomaviral DNA in the sarcoid of the donkey (poster). pp 135–137 *in*: Fielding and Pearson (eds), Donkeys, mules and horses in tropical agricultural development. Proceedings of colloquium held 3–6 September 1990, Edinburgh, Scotland. Centre for Tropical Veterinary Medecine, University of Edinburgh, UK.

Reid, S.W.J., Mohammed, H.O. 1996. Survival analysis approach to risk factors associated with hyperlipemia in donkeys. *Journal of the American Veterinary Medical Association* 209(**8**)**:** 1449–1452.

Rossel, S., Marshall, F., Joris P., Pilgram, T., Adams, M. and O'Connor, D., 2008. Domestication of the donkey: Timing, processes, and indicators. Dolores R. Piperno, Smithsonian Institution, Washington, DC, and Balboa, Panama.

Sambraus, H.H. 1994. Guide des animaux d'élevage (250 races en couleurs). Eugen Ulmer eds. 304p. (pp. 272-273).

Sarr, J., Diop, M., and Cissokho, S., 1988. La peste équine africaine au Sénégal: état de l'immunité naturelle et/ou acquise des chevaux autour des foyers récents. *Revue Elev. Med. Vét. Pays trop.*, 1988, 41 (**3**)**:** 243-246.

Slingerland M A 1989. Selection of animals for work in sub-Saharan Africa: - research at the ICRISAT Sahelian Centre. *In*: Draught Animals in Rural Development (eds. D. Hoffman, J. Nari, R.J. Petheram), pp. 203 - 210. *ACIAR Proceedings Series No. 27,* Canberra, Austrialia.

Smith, D.G., Nahius, A., Archibald, R.F. 1994. A comparison of the energy requirements for work in donkeys, ponies and cattle. pp 17–22 *in*: Bakkoury M and Prentis R A (eds). Working Equines. Proceedings of the second international colloquium held 20–22 April 1994, Rabat, Morocco. Actes Éditions, Institut Agronomique et Vétérinaire Hassan II, BP 6202-Instituts, 10 101 Rabat, Morocco.

Starkey, P. 1994. Donkey utilization in sub-Saharan Africa: recent changes and apparent needs. pp. 289–302 *in*: Bakkoury M and Prentis R A (eds). Working Equines. Proceedings of the second international colloquium held 20–22 April 1994, Rabat, Morocco. Actes Éditions, Institut Agronomique et Vétérinaire Hassan II, BP 6202-Instituts, 10 101 Rabat, Morocco.

Svendsen, E. D. (ed), 1989. The Professional Handbook of the Donkey, 2nd Edition. The Donkey Sanctuary, Devon, UK.

Wambui, C., Abdulrazak, S.A., Ondiek, J.O. and Ogore, P.B. 2004. Role and management of donkeys at farm level in Nakuru district, Kenya in Fielding and Starkey eds Donkeys, people and development: A resource book for the Animal Traction Network for Eastern and Southern Africa. CTA, Wageningen, Netherlands. pp. 24-29.

Wells, D., Krecek, R.C. and Kneale, J.A. 2004. Socio-economic and health aspects of donkeys in North-West and Cape Provinces, South Africa in Fielding and Starkey eds Donkeys, people and development: A resource book for the Animal Traction Network for Eastern and Southern Africa. CTA, Wageningen, Netherlands. pp. 203-208.

Wilson R T 1990. The Donkey. In: An Introduction to Animal Husbandry in the Tropics. W.J.A. Payne. Longman. pp. 581 - 603.

Wold, A.G., Tegegne, A. and Yami, A. 2004. Research needs of donkey utilisation in Ethiopia in Fielding and Starkey eds Donkeys, people and development: A resource book for the Animal Traction Network for Eastern and Southern Africa. CTA, Wageningen, Netherlands. pp. 77-81.

Yilmaz, O. and Wilson, R.T. 2013. The domestic livestock resources of Turkey: notes on donkeys. *The Journal of Animal and Plant Sciences* 23(**2**): 651-656.

Yman, I.M., Sandberg, K. 1987. Differentiation of meat from horse, donkey and their hybrids, mule and hinny, by electrophoretic separation of albumin. *Meat Science* 21(**1**): 15–24.

Youssef M K, Dill D B and Freeland D V 1972. Energetic Cost of grade walking in man and burro, *Equus asinus*: desert and mountain. *Journal of Applied Physiology* **33:** 337 - 340.

Index

9 789354 614507

Printed at Repro India Ltd.